Karl Heinz Koch
Process Analytical Chemistry

Springer

Berlin
Heidelberg
New York
Barcelona
Hong Kong
London
Milan
Paris
Singapore
Tokyo

Karl Heinz Koch

Process Analytical Chemistry
Control, Optimization, Quality, Economy

With 97 figures

Springer

Professor Dr. Karl Heinz Koch
Auf dem Mühlenhofe 41
D-44267 Dortmund

Title of the German edition: Industrielle Prozeßanalytik, Springer-Verlag
Berlin Heidelberg 1997

ISBN 3-540-65337-6 Springer-Verlag Berlin Heidelberg New York

Library of Congress Cataloging-in-Publication Data
Koch, Karl Heinz:
Process Analytical Chemistry: Control, Optimization, Quality, Economy / Karl Heinz Koch.–Berlin;
Heidelberg; New York; Barcelona; Hong Kong; London; Milan; Paris; Singapore; Tokyo:
Springer, 1999
    ISBN 3-540-65337-6
    Dt.Ausg.u.d.T.: Industrielle Prozeßanalytik, 1997

© Springer-Verlag Berlin Heidelberg 1999
Printed in Germany

Cover-Design: Design & Production, Heidelberg
Typesetting: MEDIO GmbH, Berlin
SPIN: 10687202   52/3020 - 5 4 3 2 1 0 - Printed on acid-free paper

Preface

Industrial process analytics is, like the whole field of analytical chemistry, an integral and essential part of every industrial company based on chemical reactions. It provides decision aids in the series of process steps, so its results have decisive technical, economic and ecological effects. Therefore, this part of analytics must hence be included in the teaching of modern "applied" analytical chemistry. Only in this way can the student, particularly the advanced one, recognise the real relevance of this scientific matter, and learn the way the methods of analytics are applied for solutions to problems in actual practice: a discrepancy between university training and working life realities is avoided and the variety in "instrumental" (physical) analytics becomes more comprehensible.

This book is therefore aimed, on the one hand, at advanced students of analytical chemistry, chemical engineering (chemical process technology), material sciences and, on the other hand, at the analytical practitioner, the chemical engineer and the process engineer who requires information about the possibilities and methods of process analytics and knowledge about their efficiency, e.g. with regard to industrial process optimization. In this case, the chemical engineers and chemical engineering technicians working in small and medium-sized companies are also included, these having to solve process analytical problems without an experienced analyst by their sides. For this purpose, lists of suppliers added to particular chapters, which naturally cannot make any claims to completeness, are intended to be helpful, and can make practical problem solving easier. For the advanced students interested in analytics this volume is intended to impart insights into the necessity for the development and the application of analytical methods in industrial practice, and it can possibly serve the purpose of orientation for advanced studies.

This presentation of industrial process analytics is supposed to be a supplement to proven text books and manuals and is made in the knowledge that more recent books in the field of industrial analytics are scarce.

For the description of the contents of the first chapters, divided according to the aggregate states, the scientific basis of the methods is briefly presented and after that the state-of-the-art is described by examples. The references annexed to each chapter are intended to provide more profound access to the matter. The publications cited can provide further help.

In some chapters examples for the interaction and the mutual dependence and reciprocal influencing of various disciplines are described. In this way, for example, the future analyst or chemical engineer is directly acquainted with the necessity of an interdisciplinary dialogue.

In a concluding chapter the integration of quality assurance – a concept which in connection with the European Market has become of considerable importance for the whole industry and for the analytics – into process engineering and process analytics is dealt with.

Finally it should be mentioned that, when dealing with particular sub-areas, the scientific basis of the analytical methods is generally presupposed.

For stimulating discussions and helpful indications the undersigned would like to thank most sincerely Prof. Dr. Manfred Grasserbauer. Furthermore, special thanks to Dr. J. Flock, Dortmund, for preparing numerous figures. Finally, many thanks to the Springer-Verlag for the unbroken realization of this project without any problem.

Dortmund/Vienna, 1999   K.H. Koch

Table of Contents

1 Introduction

1.1
Fields of Application of Industrial Analytics

First of all, the term of (chemical) "analytics" should be briefly explained in order to be able to classify that special part of analytical chemistry [1], the *industrial process analytics (process analytical chemistry)*, which will be treated here. By "analytics" we understand the obtaining of information not only on the qualitative and/or quantitative composition, but also on the geometrical structure of substances [1], including the sampling and the preparation of the material to be investigated and the very difficult and time-consuming evaluation of measured results [2] (chemometrics [3]). This processing of analytical data includes, in special cases, the process step of data reduction in order to obtain a plausible and directly comprehensible result. From this characterization it follows that analytics goes far beyond the field of traditional analytical chemistry [4, 5].

The analytical *result* is, in certain circumstances, not only useful for the industrial client (process engineer) or the researcher but at the same time it is of importance for the consumer (items of everyday use), the legislator [6] (cf., e.g., the German Chemicals Act "Chemikaliengesetz"; see Sect. 9.2.2) and/or the media as representatives of the public interest (environmental relevance) [7]. In order to reach the analytical goal a strategy is required which has its starting point in the description of the object and after that defines the methods for

---

1 General *definition*: "Analytical chemistry" is the science of the synoptical micro- and/or macrological observation and informational processing of the material-related and reagent-dependent signals from the chemical, physical or biochemical reactions between sample and reagent which leads to the clarification of the substance.

Definition by K. Cammann *(Competition: "Analytical chemistry – today's definition and interpretation", 1992)*: Analytical Chemistry is defined as the self-reliant, chemical sub-discipline which develops and delivers appropriate methods and tools to gain information on the composition and structure of matter, especially concerning type, number, energetic state and geometrical arrangement of atoms and molecules in general or within any given sample volume.

Definition of the Working Party of Analytical Chemistry (WPAC) of the FECS (EUROANALYSIS VIII, Edinburgh (UK), 1993: "Analytical chemistry is a scientific discipline which develops and applies methods, instruments and strategies to obtain information on the composition and nature of matter in space and time."

Object:

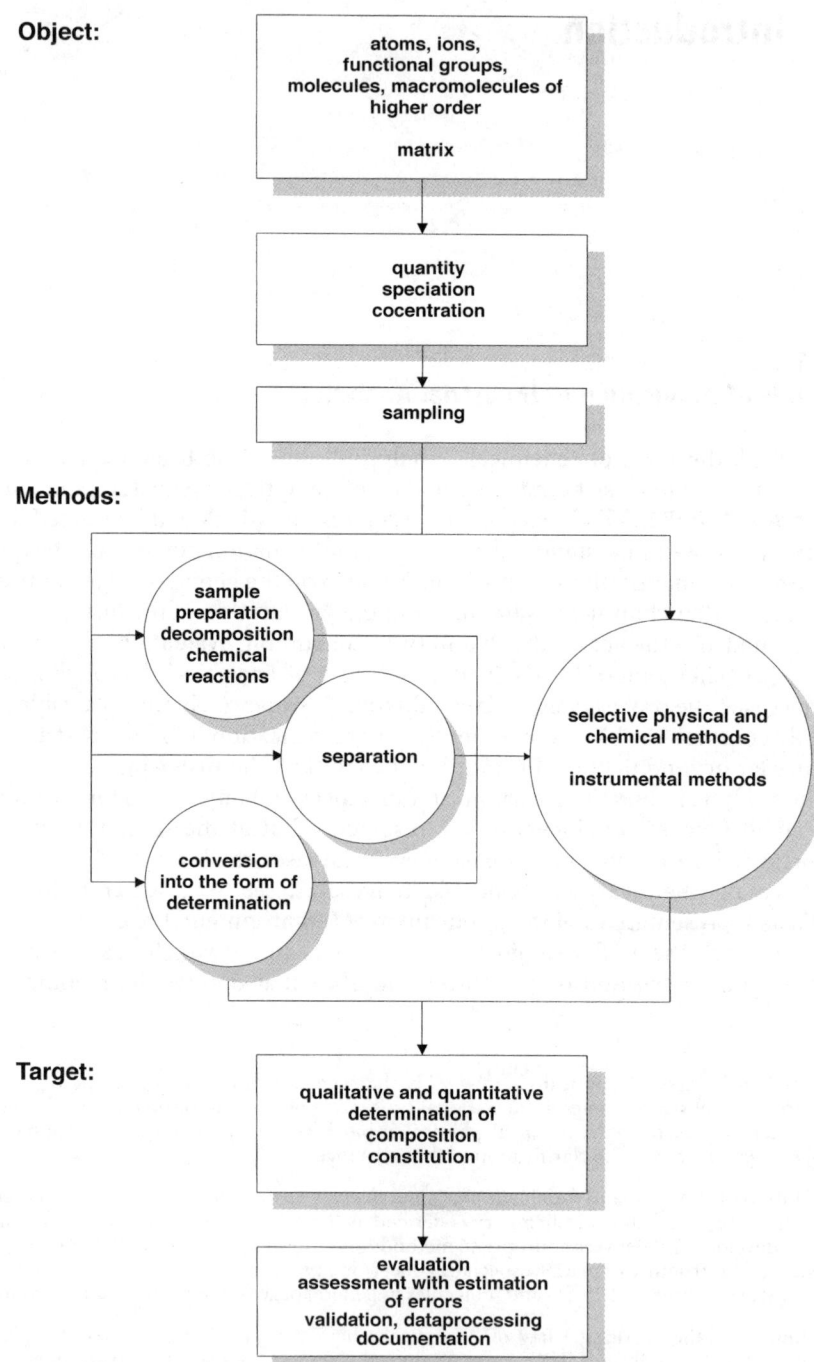

Fig. 1.1. Strategy of analytics

reaching the goal (Fig. 1.1). With regard to the methods it must be noted that modern analytics is characterized by a variety which, by a combination of methods, permits the solution of the most complicated questions. As an example of the temporal development of this broadening of analytical methodology, attention should be drawn to the steel industry as one of the basic materials industries (Fig. 1.2). While in 1950 only purely chemical methods were used, the widespread introduction of atomic spectroscopy started in 1960; in 1970 phase analytics was already of great significance and the application of gas chromatography as well as of infrared spectroscopy began; since 1980 surface analytics became increasingly important. The factors decisive for these developments will be explained at a later stage.

The title of this book *"Process Analytical Chemistry"* already outlines the programme for the subject field [8] to be dealt with and contains a subject-oriented claim which is to be defined below. According to it, process analytical chemistry [9] is to be understood as part of the instrumental analytics used in process engineering which means the application of multi-element and/or multi-method concepts [10]. This field of process analytics – or rather "chemical process analytics" – is thus demarcated as against the process-accompanying measurement of physical variables such as temperature, pressure, viscosity, etc., which only under certain conditions can be regarded as process analytics. The (chemical) process analytics covers discontinuously and continuously working methods, whereby in-line and on-line procedures [2] are promoted and developed more and more [11]. The latter are gaining increasingly in technical and economic significance, in the course of which, during the development phase, considerable material-related problems frequently have to be solved [12].

The field of application of *industrial analytics* as a whole[13] naturally goes far beyond the area of process analytics in the stricter sense already outlined [14,15]. Complementary to the purely process-accompanying and product-describing investigations comes the analysis of

- raw materials of the most varied type (possibly including samples from ore prospecting or raw material production),
- by-products of various process stages,
- competitive products in various markets,
- auxiliary materials such as boiler feed-water, water fit for industrial use, lubricants, fuels, gaseous, liquid and solid fuels, construction and painting materials,
- waste gases and waste water including their assessment with regard to environmental matters and legal regulations,

---

2 in-line = investigation in the production flow (without sample taking)
  on-line = investigation of partial quantities continuously sampled and analysed
  off-line = investigation of samples discontinuously sampled and analysed without direct (automatic) linking to the process
  at-line = quick testing near the process

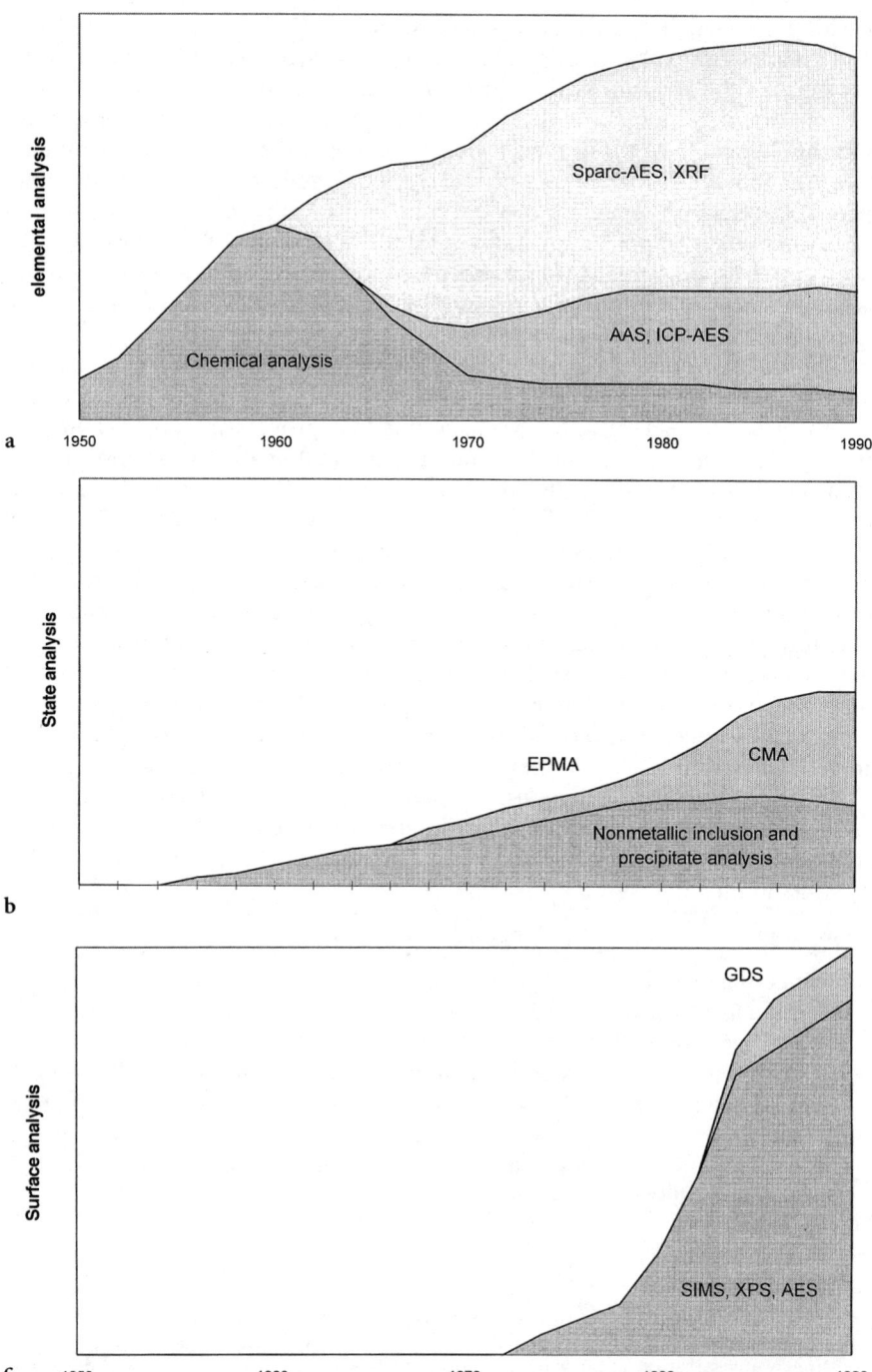

Fig. 1.2 a–c. Historic development of chemical analytics in the steel industry (since 1950)

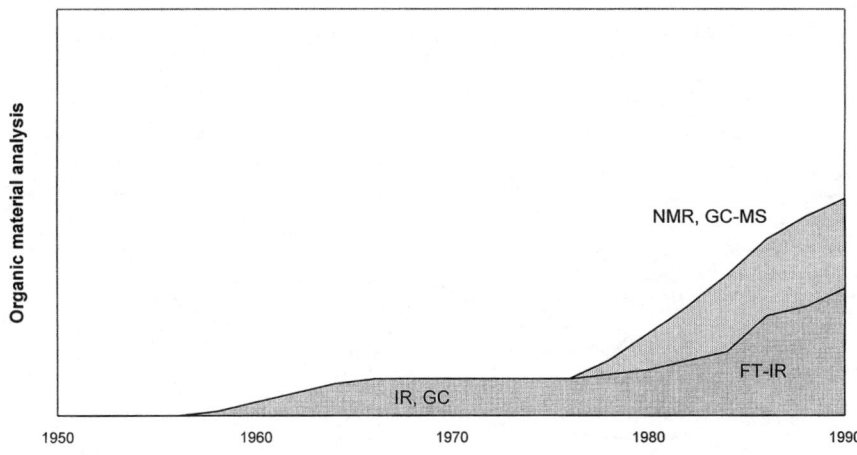

Fig. 1.2 d. Historic development of chemical analytics in the steel industry (since 1950)

- detergents and cleansing agents for the most varied purposes (cleaning of machine parts, plants, workshops, laundries, etc.),
- samples from the fields of ergonomics and industrial medicine.

(This list cannot make any claim to completeness due to the difference in the analytical requirements in the individual sectors of industry).

The levels of investment and efforts for specific research projects by the most important industrial branches to safeguard their future business differ widely and depend on a number of technical and economic factors. In any case, these future-safeguarding measures include, to a not inconsiderable extent, development and application of analytical methods. The same applies to competition between various materials which is characterized by a large range of substitution tendencies (Fig. 1.3). Here, too, analytics plays an important part in the characterization of conventional and newly developed materials as well as the description of their chemical properties.

The activities mentioned comprise, in addition, problem-related research and development work and the training of staff and of new young personnel (professional training of chemical laboratory assistants, preparation of application-orientated dissertations and theses).

Considerable technical and economic importance [11] is attached to *process analytical chemistry*, as will be shown by a number of impressive examples. The technical importance lies partly in the fact that this field of analytics enables the description and control of technical processes and the characterization of the products. The economic aspect consists of the creation of preconditions for cost minimization of process technology. Moreover, this area of analytics delivers a considerable contribution to quality control of products which will be looked at more closely elsewhere.

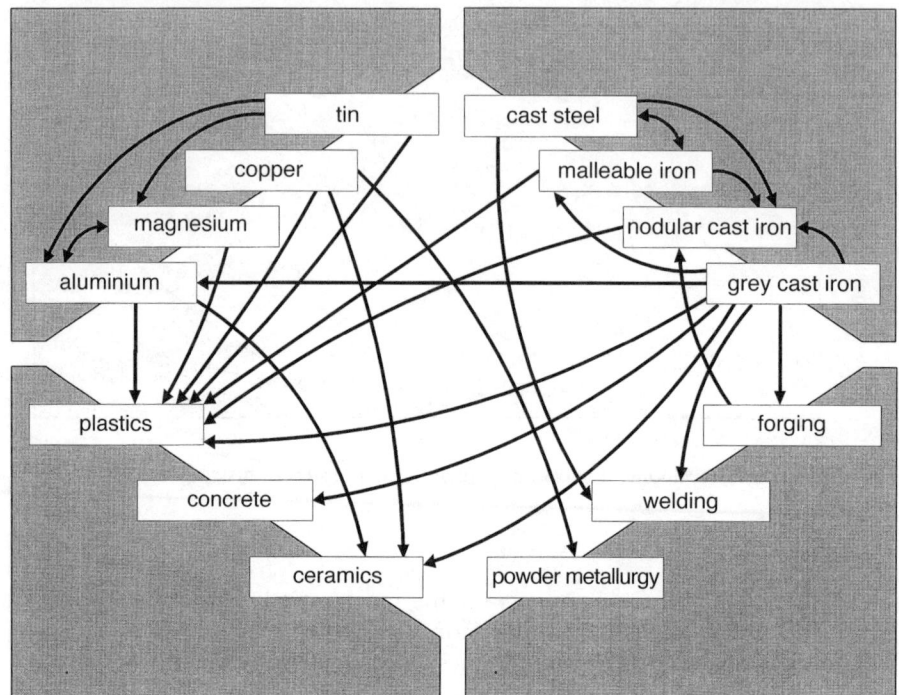

Fig. 1.3. Substitution tendencies of materials

Also, with regard to the *automation of industrial processes*, an outstanding · role is played by analytics [17] as it is, in many cases, only possible to create the necessary preconditions for successful process control with its help [10, 18]. This fact frequently makes it necessary first to automate the investigation procedure itself [19, 20]. Accordingly, we talk of automation *within* analytical chemistry and *with* analytical chemistry [21–27].

Thus the terms "automation" and "automatization" are referred to. One definition of automation reads "use of mechanical and instrumental devices in order to replace human work and abilities for the performance of a given process, partially or completely"; such systems are controlled by the feedback of information in order to enable self-controlling or self-adjusting equipment. The higher the degree of automation thus reached the more important are diagnoses of accidents and the remedying of disturbances. The skills required characterize the human being far more than any technology. A high availability of a plant thus necessitates the qualified *human*. He is the only one who is able to react flexibly and to improvise. It thus follows that in future, too, it will not be possible to dispense with humans for the control of technical processes.

Terms just used such as control, regulation and automation require, in view of the following description of various process sequences and for a better understanding of the overall problem complex, a brief explanation. Therefore there follows a selection of the most important terms and their contents [28].

Open loop control	conversion of information received with the help of measuring instruments by humans (see Fig. 1.4).
Closed loop control	conversion of information received from instruments into control variables for the process. The human has a supervisory function.
Instrument	device which converts properties and states into phenomenologically useable information (measuring, calculating or advising of an existing state). Instrumentation: use of instruments.
Mechanism	arrangement of moveable objects with a defined effect.
Machine	device with mechanisms for the repeated performance of a predefined task.
Mechanization	use of machines (in order to replace work by humans). To mechanize: to use machines.
Automatic machine	device consisting of mechanisms and instruments which constitutes a closed system (feedback of information).
Automation	use of automatic machines.
Automatization	(not commonly used) use of automatic machines with the involvement of cybernetics.

1.2
On the History of Process Analytical Chemistry

The origins of process analytical chemistry can be seen in operating and raw materials control as it came to be applied on a wide scale early this century. It is not uninteresting to have a brief look at the historical development in order to be better able to appreciate the level reached by now and the importance of this branch of analytics. In many cases it can be seen that the *"chemical/technical analysis"* – a term which was once usual for this technically oriented part of analytics – can be regarded as a motive for the *whole* analytical development. Many methods have been invented as "technical" procedures which only achieved scientific depth and recognition in the further process (Table 1.1).

Volumetric analysis proposed by Descroizilles in 1795 was initially used *only* for the control and value determination of technical products, namely acids and alkalis. Studies on chlorimetry (1824), alkalimetry (1828) and for silver determination (1832) by J.L. Gay-Lussac and the permanganate method for determining iron by F. Margueritte (1846) also arose from technological demands. It was only work by Bunsen (1853), Schwarz (1850) and Mohr (1855) that finally led to the scientific justification and recognition of these titrimetric methods.

Explanation:

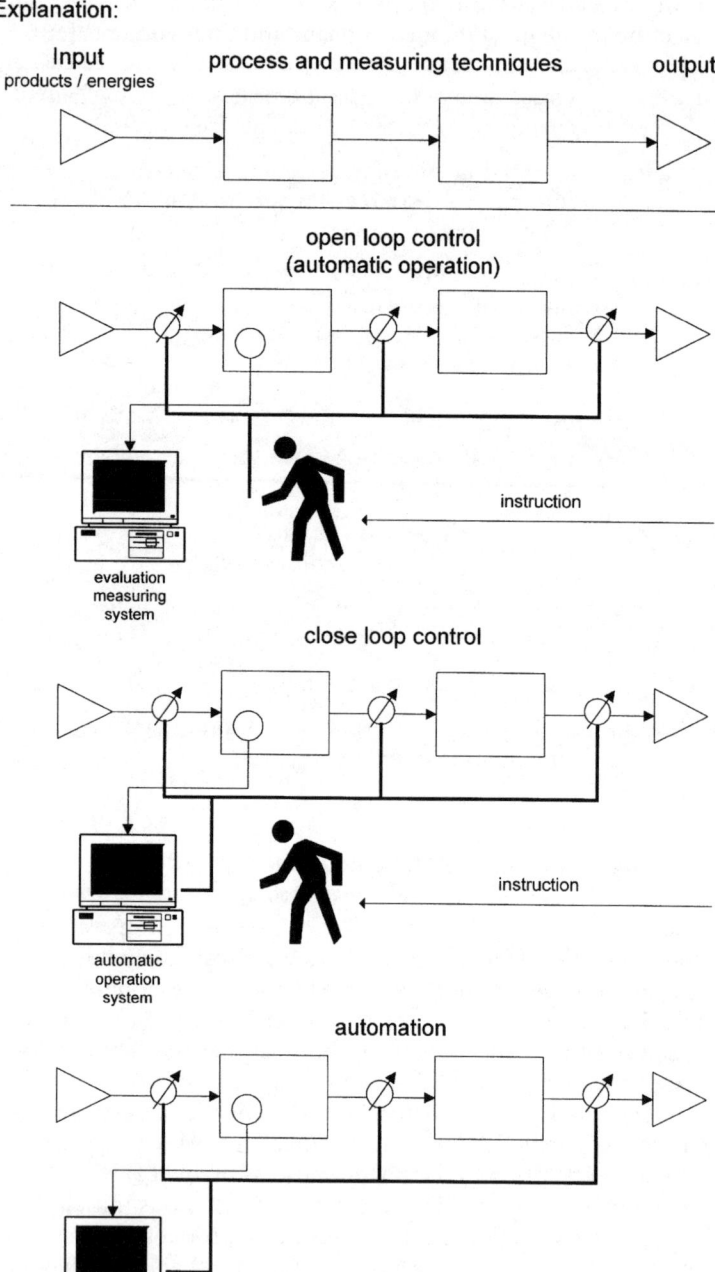

Fig. 1.4. Principle of control, regulation and automation

Table 1.1. Examples in the history of analytical chemistry

Year	Author	Methodology
		Titrimetry
1795	Descroizilles	Titrimetric investigation of technical acids and alkalis
1824	Gay-Lussac	Chlorimetry
1828	Gay-Lussac	Alkalimetry
1832	Gay-Lussay	Titrimetric determination of silver
1846	Margueritte	Permanganate method for the determination of iron
1850	Schwarz	
1853	Bunsen	Scientific justification and recognition of titrimetry
1855	Mohr	
		Emission spectroscopy
1867	Lielegg	Spectral observation of the Bessemer flame phenomena (Air refining process for steel production)
1936	Thanheiser und Heyes	Emission spektral analysis of steels by direct photoelectric determination of individual elements
1955		Begining of the application of emission spektrometers in the european steel industry

Early this century new editions of the first comprehensive descriptions of "chemical/technical analysis methods" [29, 30] were produced, which already contained the subdivision of the subject fields into raw material and operating and final product control, still of interest today.

As a historical example the control of sulphuric acid production by the lead chamber process could be mentioned [30]. The investigation of the *raw materials* in that case included the chemical analysis of pyrites, blends or elementary sulphur as well as of saltpeter and nitric acid. The "*operating control*" consisted of the regular analysis of the roaster gases, the chamber gases (SO_2, nitrous gases), the outlet gases (O_2, acid content) and the "process acid", i.e. the intermediate products obtained at various points of the process. At the end the *final product,* sulphuric acid, was analyzed according to the particular criteria required in each case.

As a remarkable historical example from the field of emission spectral analysis it should be mentioned that as early as 1867 the Austrian secondary school teacher A. Lielegg proposed spectral observations of the flame phenomena emitted by Bessemer converters (air refining process for steel production invented by Henry Bessemer in 1855) for the improvement of process monitoring [31]. The papers which appeared at that time already contained all the main criteria which in the recent past led to the comprehensive use of emission spectrometry in metallurgy.

However, it was about 70 years later when, in 1936, Thanheiser and Heyes at the Kaiser-Wilhelm-Institut für Eisenforschung (nowadays Max-Planck-Institut) in Düsseldorf succeeded for the first time in achieving *direct photoelectric* determination of manganese and chromium in steels [32].

This development was continued in the USA in the 1940s. The first emission spectrometers with spark excitation ("quantometer"®) were produced, and in 1955 the use of the first imported devices in the German steel industry started. Since then the production of steel has been directly linked with emission spectrometry and is dependent on it [33]. At this point it must also be mentioned that after more than 130 years Lielegg's actual goal, the direct monitoring of a metallurgical process (without the sampling step), has not yet satisfactorily been achieved (for further details see below).

1.3
Economic Importance of Process Analytical Chemistry

Increasing quality demands, the constant search after processes of high effectiveness and higher productivity, optimization of known processes and environment legislation present the mainsprings for the constant growth of the market, particularly with analytical on-line systems. While the markets in the USA and Europe will develop continuously, but moderately, until the year 2000, a large break-through in the field of process analytics will approach the Asiatic-Pacific region in the next years, so that from these states the largest inquiry push will come. The annual growth instalments will thereby amount to an average of 6% as a study of the market research institute Frost & Sullivan, London, shows. After that, on-line analytics, which experienced its first large upswing in the 1980s, will now expand its market share considerably.

While in North America applications in the field of the production of purest water will exert the main influence on the increase of on-line analytics, in Europe water and waste water monitoring as well as pharmaceutical and paper production control are important fields of application. Furthermore, a significant mainspring for the growth of the analytics market will naturally be delivered by the chemical industry. Strong impulses on the world market will come from the Japanese industry because of the increasing process control in numerous branches of industry.

In this situation the world market of analytical equipment producers always becomes more contest-intensive. The wave of acquisitions and fusions of leading producers will continue, whereby national boundaries become more and more unimportant.

Besides the industrial management aspect linked with the particular operating result of a production plant or the whole company, process analytics – nationally and internationally speaking – also has an interesting staff management component as the following examples of the metallurgical and chemical industry show.

In the steel industry world-wide approximately 10,000 emission spectrometers are installed which – including the specific costs for the laboratories, the devices for sampling, for the preparation, handling and transport of the samples as well as for data processing – represent roughly invested capital [34] of 10,000 × 600,000.00 DM = 6 billion DM. (The industrially used X-ray fluorescence, atomic absorption and ICP spectrometers, the numbers of which are of the same order of magnitude, were not included in this calculation!)

If one ties the number of these instruments with the persons working with them the following picture is obtained.

As the majority of spectrometers are used for production control in multi-shift operation an average value per item of equipment (sample preparation, operation, supervision, maintenance) in the case of three-shift operation (including provision of replacements) can be expected as being four persons, i.e. alone with emission spectrometry in the steel industry, world-wide 4 × 10,000 = 40,000 persons(!) are employed.

Thereby, this number is of the same order of magnitude as the workforce figures of international companies with annual turnovers of the order of 10–20 billion DM.

The second example concerns the application of chromatography in the chemical industry. It is interesting to ascertain that, in the normal analytical laboratory in this branch of industry, about two thirds of all methods applied are based on chromatographic principles, while spectroscopic methods comprise about a quarter so that the remaining analytical procedures represent less than 10%. The economic importance of chromatography illuminates the fact that world-wide about 100,000 liquid chromatographic devices are used in daily routine operations.

1.4
Presentation and Organizational Forms of Process Analytical Chemistry

For the description of methods and fields of application of (industrial) process analytics there are method- and material-oriented ways of looking at them.

Neither of these types of description is useable alone as a classifying principle because of the complexity of this subject area. Therefore, a combination of the two will be tried below. To provide a general idea of the methodological basis the presentation is made method-oriented, being substance-oriented in the stratified examples of operating practice. As a further classifying principle, division into the three aggregate states – gaseous, liquid and solid – of the particular process substances will be carried out.

The individual sections will start with the treatment of the methods of sampling and sample preparation, then cover the analytical methods, and end with examples of industrial practice. Thereafter, in special chapters the close interlinking of process engineering and analytics, their interdependence and their mutual influences will be shown. The final chapters deal with questions of quality inspection and assurance of test results, and provides a look at future-oriented lines of research and development.

Before dealing with the individual chapters of process analytics, the organization of analytical laboratories in industry should be examined [13]. While in large companies in the chemical industry analytical activities are frequently organized on a decentralized basis, the organization of analytics in works of the metal industry or in plastics processing is centralized [35].

Centralization of the whole of chemical and analytical activities in one laboratory area with its specially trained and skilled staff has not only the advantage of optimal selection of personnel for a given problem, but also the expenditure on expensive analytical equipment and their maintenance permits the most favourable efficiency [36]. Further specialization and instrumentation in analytics should, to be consistent, have to urge further centralization of analytical laboratories. In the future, the further development and application of sensor technology may lead – in many cases – to decentralized works practice.

1.5
Importance of Sampling

The description of procedures in chemical process analytics, for which – apart from in-line procedures – the process step of sampling is of fundamental importance, should be preceded by the explanation of some important terms and definitions. That appears to be absolutely essential in order to avoid misunderstandings in the interpretation of results or goods traffic by the use of uniform terminology.

For that purpose it must be explained first of all what is to be understood in chemical analytics by "sampling" [37]. Here, "sampling" means the removal of increments of a bulk for test or investigation purposes, with all the properties of this part (sample) having to match those of the main mass. The sample "taken" must be representative for the sampled material [38]. For the sampling of the various substances there are a number of directives (standards) and methods for achieving the stated goal.

The importance of subject- and substance-relevant sampling becomes clear by the example of melt control in a steelmaking plant. In this case, a sample of about 80 g in mass must be characteristic for a melt of 200–300 tonnes of steel (mass ratio $1:2 \times 10^6$). If analysis is done by means of spark emission spectrometry, only about 1 mg of this sample evaporates (mass ratio $1:10^{11}$!), of which, however, only a fraction is excited and becomes spectro-analytically effective. Thus the demand for the homogeneity of a sample with regard to a reliable analytical result becomes clear.

A reliable sampling of large bulk materials can only be ensured by means of automatic sampling equipment which, in certain cases, can require capital expenditure of several million DM and can be independent works plants [39]. Each sampling is closely related to the following investigations and is to be considered the first step of analysis. Therefore, the analyst must also be able to have a direct influence on sampling [40].

In connection with sampling and in view of the following analytical procedures, uniform, clearly defined terms, the aim of every standardization activity (see, e.g., the German DIN 55 350, Part 14 [41]), should be used.

The following explanations of individual terms are taken from the "Handbuch für das Eisenhüttenlaboratorium" (Manual for the iron and steel works laboratory) [42]. More comprehensive descriptions have been recorded, for the most varied subject fields, in numerous standards to which reference will be made when required. Terms occurring in excess of them will be explained in context.

Selection of definitions for sampling and sample preparation:

- *Analysis sample* (portion of analysis): sample for performing analyses, e.g. for determining the chemical composition or the phase structure. A distinction is made, e.g. between gaseous, liquid, powdery, chipped, lumpy, pressed and remelted analysis samples.
- *Effective analysis sample* (portion of determination): amount of the analysis sample which is actually taken for performing an analytical method, e.g. the weighed sample during the chemical analysis, the mass subjected to radiation during X-ray fluorescence analysis, the evaporated sample mass during emission spectrometry.
- *Increment* (portion of sample): smallest quantity of a sample material which is taken from one lot in one sampling step.
- *Homogeneity*: uniformity of a material. A material mass is homogeneous if the attribute to be checked exists to the same extent in all the divided quantities relevant for or specified for the determination of homogeneity. As a measure of homogeneity the standard deviation of the attribute in the quantity of material (lot) to be sampled can be used if the standard deviation for the test procedure is correspondingly low.
- *Laboratory sample*: sample which after partially performed or completed sample conditioning is ready for dispatching to the laboratory. It must not be identical with the analysis sample.
- *Lot*: mass of material from which a sample is to be obtained.
- *Sample* (portion of substance): a mass of material representative of the lot. The term "sample" in sampling is identical to the term "random sample" in statistics.
- *Sample material*: the material in each case gained during sampling or sample conditioning or continued material after sub-dividing the sample material.
- *Sample portion:* part separated from a product from which the analysis sample and/or the samples for other tests are obtained.
- *Sampling*: whole process from the removal of samples or increments and possibly also comminution, mixing and sub-dividing to the obtaining of a sample for the desired purpose. It consists of sample removal and sample conditioning.
- *Sample conditioning:* the crushing, mixing and sub-dividing of the sample product for the obtaining of the analysis sample or the sample for a different test.
- *Sample removal:* removal of samples or increments from the material to be tested according to a specified plan.

- *Sample preparation*: preparation of an analysis sample, or of a sample intended for a different test. This includes, e.g. in the case of combustion methods, the weighing in, in the chemical analysis the weighing in, dissolving and carrying out of reactions, and in the case of spectral analysis of solid samples the grinding or milling.
- *Sample piece*: the piece selected from the lot for the removal of samples, e.g. the selected plate or the selected rod.
- *Representativeness*: characterizing of a sample with regard to the constituent to be determined. The sample is representative of the material to be examined or of the lot when the mass percentage of that constituent in the sample matches with that in the whole material to be tested within the limits of the test accuracy. If a property has to be tested, a sample is representative if this property in the sample matches that in the whole material to be tested within the limits of test accuracy.
- *Bulk sample*: combining of all increments.
- *Partial (divided) sample*: combining of increments in groups leads to partial (divided) samples.

With regard to the international system of measures (SI) which, besides mass (in kg), includes the substance quantity (in mole) as a basic variable, it was essential for a quite definite, delimited quantum of a substance to use an adequate expression which had to be independent of any properties still to be determined, in particular of variables. Unusual but quite practicable, and similar to Anglo-Saxon language usage, this (still undefined but clearly definable) quantum of matter is termed a "portion of substance" (German standard DIN 32 629) [43]. With regard to sampling and analytical method a distinction must consequently be made between that portion of substance which is a representative part of the whole of the material to be analysed (the portion of sample) and of that part of it which is analyzed (the portion of analysis). The latter contains the amount of that to be determined (the portion of determination). It can be sub-divided during the analytical procedure into several aliquot portions of determination. Of all portions of substance, the mass or the substance quantity can be determined. The corresponding basic units are the gram and/or the mole.

The sample is, however, not only an accumulation of matter particles, but also information. It contains the "construction" plan, the exploration of which is to be performed by means of the extraction of information [44]. The "synholon" [3] sample is therefore in the centre of a scientifically well-founded work and, besides the qualitative and quantitative matter parameters, increasingly the

---

3 The philosophical term "synholon" used in connection with a "sample" is composed of the two Greek words syn (συν)=(adv.) together, at the same time or (prep.) at the same time as, provided with and holon (τὸ ὅλον)=the whole, space, main thing (ὅλοζ=whole, complete). Therefore synholon means a whole (a portion of substance) which contains all the information (attributes) which can be used for the description of this whole, such as chemical composition, form of bond of the individual elements, crystallographic structure, physical properties etc.

main emphasis is to be placed on the form parameters (form/matter reference of the sample consideration).

References

1. Kelker H, Tölg G (1994) In: Ullmann's encyclopedia of industrial chemistry. VCH, Weinheim, vol B5, p 1
2. Doerffel K, Henrion R, Henrion G (1994) In: Ullmann's encyclopedia of industrial chemistry. VCH, Weinheim, vol B5, p 35
3. Danzer K (1992) Mitt GDCh-Fachgruppe Anal Chem 4/92, M 104
4. Danzer K, Than E, Molch D, Küchler L (1987) Analytik – Systematischer Überblick, 2. Aufl, Wissenschaftl Verlagsges mbH, Stuttgart
5. Ballschmiter K (1993) Mitt GDCh-Fachgruppe Anal Chem 2/93 M 47
6. Hein H (1993) Labor 2000, LaborPraxis 1993:152
7. Hoffmann H-J (1993) Labor 2000, LaborPraxis 1993:158
8. Riebe MT, Eustace DJ (1990) Anal Chem 62:2, 65A
9. Melzer W, Jaenicke D (1980) Prozeßanalytik. In: Ullmanns Encyklopädie der technischen Chemie, 4. Aufl, Verlag Chemie, Weinheim, Bd 5:891; Hengstenberg J, Sturm B, Winkler O (Eds.) (1980) "Messen, Steuern und Regeln in der Chemischen Technik, Bd. II Messung von Stoffeigenschaften und Konzentrationen (Physikalische Analytik)", 3. Aufl., Springer-Verlag, Berlin-Heidelberg-New York; Profos P, Pfeifer T (Eds.) (1992) "Handbuch der industriellen Meßtechnik", 5. Aufl., R. Oldenbourg Verlag, München-Wien; Oesterle G (1995) "Prozeßanalytik – Grundlagen und Praxis", R. Oldenbourg Verlag, München–Wien
10. Koch KH (1987) Mikrochim Acta I:151
11. Kaiser MA, Ullman AH (1988) Anal Chem 60:823A
12. Jakobs SM, Mehta SM (1988) Int Lab (May):20
13. Leithe W (1964) Analytische Chemie in der industriellen Praxis, Methoden der Analyse in der Chemie, Bd. 2, Akadem. Verlagsgesellschaft, Frankfurt/M
14. Koch KH (1985) Fresenius' Z Anal Chem 321:1
15. Koch KH (1993) CLB Chem Lab Biotechn 44:120
16. Borman S (1987) Anal Chem 59: 14, 901A
17. Voetter H, Huyten F (1969) Fresenius' Z Anal Chem 245:11
18. Kienitz H, Kaiser R (1966) Fresenius' Z Anal Chem 222:119
19. Bartels H (1981) In: Analytiker-Taschenbuch, Bd. 2. Springer, Berlin Heidelberg New York, p 31
20. Ebel S (1987) In: Analytiker-Taschenbuch, Bd 6. Springer, Berlin Heidelberg New York
21. Malissa H (1972) Automation in und mit der Analytischen Chemie, Moderne Analytische Chemie, Bd. 5/1, Verlag der Wiener Medizinischen Akademie, Wien
22. Malissa H (1966) Fresenius' Z Anal Chem 222:100
23. Malissa H, Jellinek G (1968) Fresenius' Z Anal Chem 238:81
24. Malissa H, Jellinek G (1969) Fresenius' Z Anal Chem 247:1
25. Malissa H (1971) Fresenius' Z Anal Chem 256:7
26. Malissa H, Rendl J (1972) Fresenius' Z Anal Chem 258:363
27. Malissa H (1974) Fresenius' Z Anal Chem 271:97
28. (1968) Fresenius' Z Anal Chem 237:81
29. Neumann B (ed) (1908) Post's chemisch-technische Analyse, Handbuch der analytischen Untersuchung zur Beaufsichtigung chemischer Betriebe, für Handel und Unterricht. 3. Aufl
30. Lunge-Berl (ed) (1921/19220 Chem.-tech. Untersuchungsmethoden, Hrsg. E.   Berl, 1. und 2. Bd., 7. Aufl., Springer, Berlin
31. Lielegg A (1867) Wiener Ber. 55:II, 153–161; 56:II, 24
32. Thanheiser G, Heyes J (1937/38) Arch Eisenhüttenwes 11:31
33. Koch KH (1984) Spectrochim Acta 39 B:1 067
34. Slickers K (1993) The automatic atomic emission spectroscopy. Brühl- Universitätsdruckerei, Gießen
35. Kipsch D (1989) Neue Hütte 34:141
36. Sansoni B (1986) Fresenius' Z Anal Chem 323:535

37. (1980) Probenahme – Theorie und Praxis", Heft 36 der Schriftenreihe der GDMB, Chemie, Weinheim
38. Gy PM (1981) Aufbereitungs-Techn Nr 12:655
39. Koch W, Roeder KP, Dobner W, Huber G (1970) Thyssenforschung 2:H. 1, 6
40. Kaiser R (1980) Fresenius' Z Anal Chem 300:9
41. DIN 55 350, p 14: Concepts of quality management and statistics; concepts of sampling
42. (1987) Handbuch für das Eisenhüttenlaboratorium, Bd 5. Verlag Stahleisen, Düsseldorf
43. DIN 32 629 (1988): Portion of substance; concept, characterization
44. Malissa H (1980) Fresenius' Z Anal Chem 300:9

2 Process Analytics of Gases

2.1
Importance for Industry and Society

The importance and the necessity for the investigation of gases for the general public is plausible because of today's state of public discussion on the environment and examples include control of air impurities with regard to measures for pollution abatement, control of pit safety in the mining industries as well as of the public safety in traffic tunnels, and use of storerooms and greater areas with artificial ventilation or for the control and dosing of fumigations for pest control. The various applications of *process analytics* include production process control in the chemical industry, continuous exhaust gas analysis for the attainment of an optimum operation of power plants, turbine aggregates and rolling mill furnaces or for the control of metallurgical processes, which will be shown by the following examples.

Continuously working gas analyzers are widely used for the regulation and control of technical processes. The classic field is certainly the flue gas analysis of *firing plants* [1]: the valuation of the combustion processes is important for the warranty of an optimum utilization of fuel with the aim of a high efficiency of the plants, and thereby care of resources and a decrease of the output of pollutants. In these cases, continuous analysis of O_2, CO and CO_2 of the flue gas is performed. Additionally, the pollutant components SO_2, NO and NO_2 are measured.

A similar objective possesses the control and regulation of *industrial furnaces*. The oxidizing, neutral or reducing character of the furnace atmosphere, on which the properties of the charged material largely depend, is continuously controlled by gas analyzers. On the one hand, these analyzers shall continuously control the waste gas composition and, on the other hand, allow the evaluation of the influence of the furnace atmosphere on the charged material. For technological reasons, the oxygen content of the flue gas in a reheating furnace should, together with heat production, be kept within certain limits in order to guarantee a low oxidizing loss via forming of cinder on the surface, e.g. of a glowing rolled piece. Therefore, the oxygen content of the furnace atmosphere and thereby the fuel/air ratio must be kept within narrow limits.

In the *chemical industry*, oxygen appears very frequently as coreactant. The determination of this element with automatically operating gas analyzers is therefore of great importance for faultless operation of processes and for obtaining an optimum yield. Here, the determination of the oxygen content, e.g. of the SO_2 mixture during sulphuric acid production in order to control the air inlet for the oxidation of SO_2 to SO_3, could be mentioned. When producing ethene oxide, ethene is oxidized under pressure in presence of a catalyst. The gas mixture of ethene and air is moved in a circle. By continuous O_2 determination, the efficiency of the process is pursued and by observation of the safety aspect it is ensured that the O_2 amount in the gas mixture remains under the explosion limit. For a trace analytical task, ammonia synthesis can serve as an example. The catalysts for ammonia synthesis are very oxygen sensitive. Therefore, the O_2 content of the N_2/H_2 mixture must be kept under 50 mg/kg O_2. This process analytical task is also solved by a continuous (electrochemical) measuring method.

Pure oxygen, pure nitrogen and pure argon have attained great technical importance during the last decades. Economic production is done by the LINDE process of *air liquefaction* and subsequent rectification. The air liquefactors and purifiers are controlled by automatic gas analysers. The continuous determination of the single gas components allows one to operate the plants with optimum output and to minimize shut-off times.

The process analytical investigation of gases also plays an important role in the *steel industry*. That is valid as well for the top gas formed by the reduction of the iron ores in the blast furnace, for the converter flue gas generated by the oxidation of carbon during steelmaking as well as the gas components formed during steel degassing by vacuum treatment of molten steel. Moreover, the control of reheating furnaces in rolling mills and the regulation of inert gas atmospheres for annealing of steels are to be mentioned. This field of gas analysis will be illuminated in more detail in connection with the process analytics of steel production and processing (see Chap. 6).

The gases to be analyzed in the framework of process analytics comprise process, burnt, inert and waste gases as well as emissions [2]. For their analysis chemical and physical methods are applied, and it is interesting that apparatuses for the fast and continuous gas analysis were described and constructed more than 90 years ago [3]. At that time, in most cases, the aim was the continuous determination of the CO_2 content of burnt gases for the regulation of firing plants. The mode of operation of a burnt gas analyzer used in the early days of process analytics (system Krell-Schultze, in 1900) was based on the determination of the mass differences of a gas and an air column of the same height. The mass difference of both equally high gas columns was indicated by a sensitive micromanometer, whereby the actual level of the measuring fluid in the micromanometer could be photographically continuously recorded by means of a light source.

Besides this principle of hydrostatic pressure difference, optical principles such as measurement of the refraction exponent or the interferometry, found an early application. Today, spectroscopic, gas chromatographic, electrochemical and chemical methods are used for process analytical purposes. Additionally, the measurement of physical effects and properties, such as density, viscosity, heat conductivity or paramagnetism are applied.

During the last years, an impetuous development of computer-controlled gas analysis devices and systems has taken place [4]. The use of microprocessors in a gas analyzer simplifies the realization of the partially complicated courses of investigation. The control of the system periphery, such as sampling probes, filters, coolers, heated pipes and pumps, by a computer leads to process-oriented installations, which can be operated with high reliability and low maintenance expenditure. The continuous measuring of low concentrations with the necessary long-time stability becomes possible by automatic recalibration of the system. Devices equipped with special sensors are, for example, in a position to recognize a break-down in the hardware and to supply reliable fault diagnoses by logical combination of several pieces of information. By processing different signals, optimized signals can be provided for process control. The high flexibility of the computer-aided gas analysis of today offers a demand-oriented adaptation to operational requirements.

Recalibration, linearization and testing of temperature-dependence, measured value resolution and long-time stability as well as documentation of instrumental parameters can all be realized automatically in the case of computer-aided systems. Combinations of several analyzers or their communication with a central computer are enabled by means of serial interfaces. By use of modems, computer-aided systems are also cross-linkable over long distances. Thereby, on the one hand, measured values can be supplied to a control room or chief office located far away and, in the case of break-downs, long distance diagnoses can be performed, which can lead to a rapid elimination of defects and can possibly avoid the supply of service personnel.

2.2
Sampling of Gases

In comparison to the states of aggregation "solid" and "liquid", sampling of gases is rather simple (see, for example, German Standard DIN 51 835) [5]. It must essentially be considered that the removal takes place with laminar flow and that wall effects in the pipes passed through are excluded. If the gas contains dust or aerosols, they have to be separated only if the composition of the pure gas phase is of interest. The sampling takes place discontinuously or continuously from the gas flow by means of probes.

For process analytics, the continuous-automatic procedures have, of course, attained outstanding importance. A multiplicity of processes in the chemical industry (see Sect. 2.1) is controlled and regulated by the continuous analysis of gases (process analyzers). The results obtained serve not only for the optimum guidance of process but also – if necessary – for the control of safety and official emission orders. The continuous sampling either takes place by a branch pipe-line or a bypass system (Fig. 2.1). In each case, a dead time (time between sampling and analysis) as short as possible has to be striven for. Therefore the bypass line is often conducted into the apparatus room directly to the analytical device.

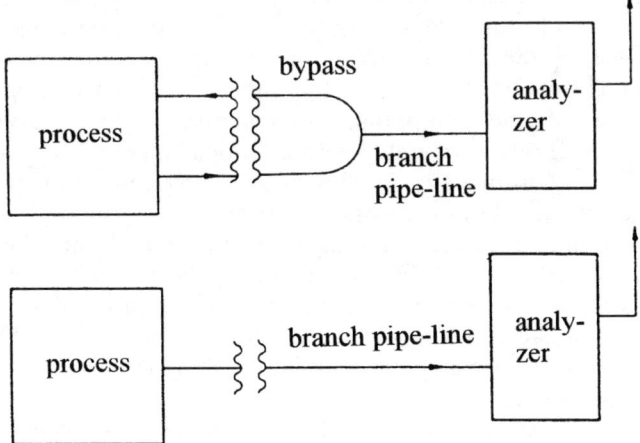

Fig. 2.1. Principle of gas sampling

Fig. 2.2.
Scheme of a gas sampling
system (by permission of
VCH Verlagsgesellschaft
mbH, Weinheim)

a = sampling probe
b = shut-off valve
c = pressure reducing
    valve
d = sampling line
e = conveying pump
f = filter
g = regulating valve
h = flowmeter

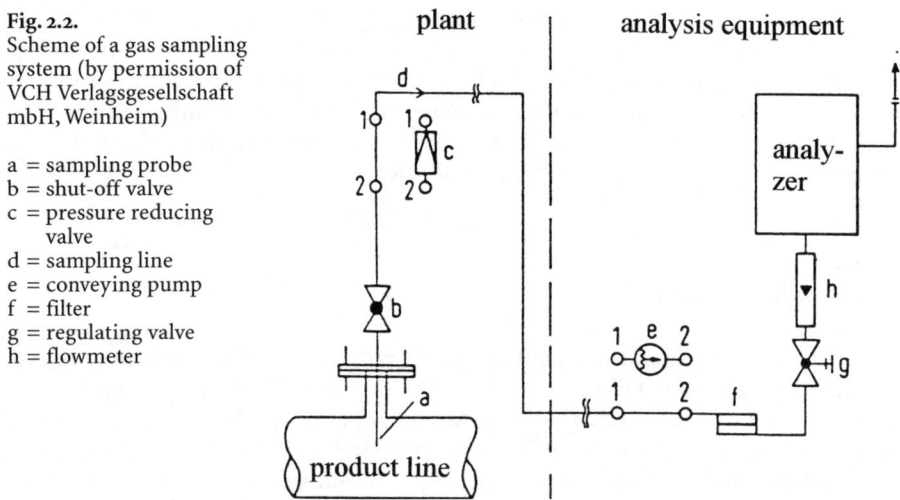

The technique of gas sampling is explained by the illustrated scheme of a simple sampling system (Fig. 2.2). The sample flow is removed by a probe out of a
product line or a vessel, respectively, and is conducted to the analyzer by a branch
pipe-line. The sampling probe is connected by a flange with the connecting piece
on the pipe-line or the vessel, respectively [2]. The use of a probe results in short
dead times because of its relatively small volume and impedes the penetration of
liquids and solid materials, which possibly go layer-like along the pipe or vessel
walls. If the gas pressure is greater than 50 kPa the shut-off valve (see Fig. 2.2) is
followed by a pressure reducing valve, and if the gas pressure is lower than 10 kPa
a conveying pump (usually membrane pump or membrane compressor) is used

before the gas is transported via a sampling line to the analyzer. The reduction of the gas pressure possibly necessary is normally arranged near the sampling place in order to keep the transportation speed through the sampling line high, and thereby keep the dead time as short as possible. Conveying pumps are usually placed near the analyzer in order to enable a common maintenance of the pump and the analytical equipment. The flow through the analyzer is regulated by a combination of needle valve and flowmeter. A filter usually protects the analyzer from contaminations. This system has to be modified in case of impeded measuring conditions (solid materials, condensable components).

2.3
Analysis of Gases

2.3.1
Measurement of Physical Effects

2.3.1.1
Paramagnetism

Gases with molecules which contain unpaired electrons are paramagnetic. Their magnetic susceptibility is positive, i.e. they are pulled into a magnetic field. To this group of gases belong O_2, NO, NO_2, ClO_2 and ClO_3, and of these gases oxygen shows the strongest paramagnetism. Therefore, this principle is almost exclusively applied for the measurement of the oxygen.

There are 3 different measuring methods for the determination of oxygen because of its magnetic properties: The thermomagnetic, the magnetochemical, and the magnetopneumatic method [6]. The measuring sensitivity for paramagnetic gases depends on the pressure and the temperature (CURIE's law). Beside these physical influences, the measurement is affected by alterations of the composition of that gas accompanying the oxygen. Therefore, measures have to be taken in order to correct these influences.

Fig. 2.3.
Example for the principle of a paramagnetic oxygen analyzer (Hartmann & Braun AG)

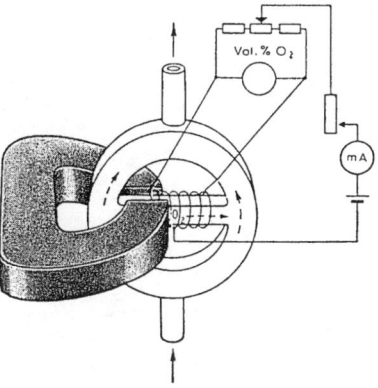

The ring-chamber apparatus based on the thermomagnetic method is widely used in process analytical practice (Fig. 2.3). In this case, the gas sample is guided through a ring-shaped chamber with a thin cross link. This connecting tube is provided with an external heatable wire winding of which one half is exposed to a strong magnetic field. Both halves are parts of a Wheatstone bridge, by which a possible temperature difference is recorded as a difference of electric resistances. If the gas sample flowing through the ring-chamber is free of oxygen, there will be no gas flow in the connecting tube, the two halves of the wire winding will have the same temperature and the bridge will be currentless. However, if the gas contains oxygen, it will be pulled into the magnetic field under preference of cold molecules (CURIE effect). The winding half lying in the magnetic field is cooled down more strongly by the local gas flow, so that the bridge indicates a current proportional to the O_2 concentration. Such devices are applicable for oxygen contents of 0.1–100 vol. %.

2.3.1.2
Heat Conductivity

The heat conductivity is also used as a quantity to be measured for gas concentrations. The heat conductivity (μW/cm K) of matter is defined by the quantity of heat Q (μWs) which flows per second with a temperature drop of 1 K/cm through the surface of 1 cm². In the measuring technique, the relative heat conductivity is usually indicated, where that of air is equal to 100. The heat conductivity is strongly temperature dependent; it increases linearly according to the kinetic gas theory for ideal gases with the middle molecular velocity. Real gases show a very much stronger increase of their heat conductivity with temperature due to the attraction of the molecules.

For the measurement of hydrogen or of impurities of other gases in hydrogen this method is especially suitable because of its high heat conductivity. Such

Fig. 2.4.
Principle of a heat conductivity detector

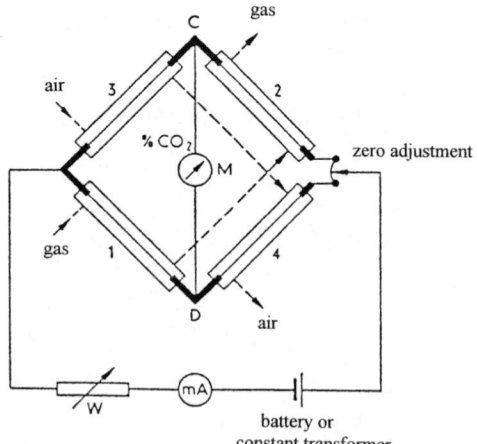

Table 2.1. Examples of application of heat conductivity measurements of gases

Analyte	Matrix	Technical process
Argon	Oxygen	Air separation
Hydrogen	Argon	Air separation
Nitrogen	Argon	Air separation
Carbon dioxide	Air	Gas generator control
Hydrogen	Carbon dioxide	Gas generator control
Hydrogen	Air	Gas generator control
Hydrogen	Blast furnace top gas	Blast furnace process
Carbon dioxide	Burnt gas	Gas-fired furnaces
Hydrogen	Inert gas	Inert gas annealing plants
Hydrogen	Air	Air control
Propane	Air	Inert gas production

instruments for process analytical purposes measure the relative heat conductivity of a gas mixture in comparison to a reference gas. For this purpose, four thin wires of platinum are stretched in narrow borings of a thermostatic metal block, which form a Wheatstone bridge (Fig. 2.4). The four bridge branches are fed with a constant current; through two alternate branches flows the gas to be measured, through the two others the reference gas [6]. The temperature of the platinum wires and thereby their resistance is essentially determined – when feed current and block temperature are constant – by the heat conductivity of the gas surrounding them.

By empirical calibration with known gas mixtures the desired measuring ranges are adjusted and the non-linearity of the measuring method, as well as influences caused constructively, e.g. by heat convection, are taken into account.

The heat conductivity method is normally restricted to binary mixtures. Table 2.1 shows some examples of application.

2.3.1.3
Heat Change

Besides the measurement of heat conductivity, the heat effect of catalytic reactions is also used as a measuring principle [6]. The heat changes occurring with the catalytic oxidation on contact can be directly determined according to changes of the resistance of measuring wires. "*Gas sensors*" based on this principle were first applied to the analysis of gas products, e.g. CO and H_2 in burnt gases, and to the control of processes. Meanwhile, catalytic "heat change gas sensors" have also been introduced into environmental analytics. Here, they are used for the quantitative determination of gaseous components in air [7]. As in this case, low trace contents often have to be determined, and great things are expected from these gas sensors regarding high sensitivity and selectivity, together with a continuous mode of operation, a high efficiency and low expenditure for maintenance.

Fig. 2.5.
Schematic structure
of a catalytic heat
change sensor

1 = Sintered metal
      disk
2 = Detector pearl
3 = Gas to be analyzed
4 = Compensator pearl

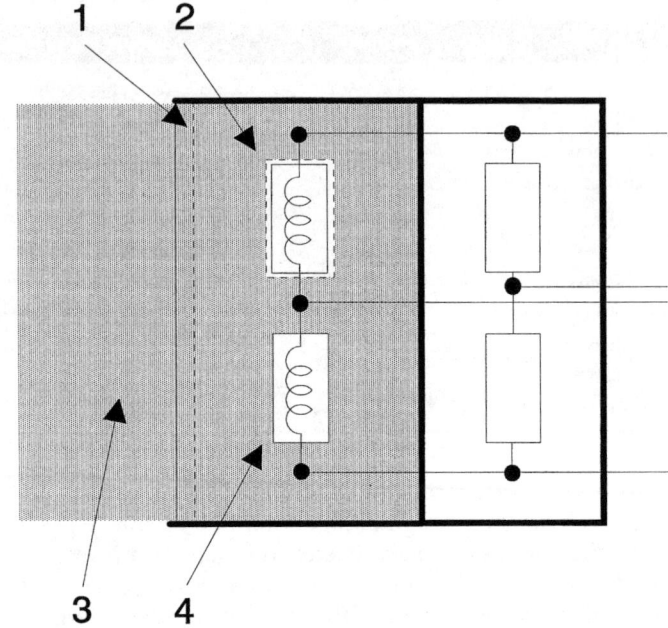

The principle of such sensors [8] is that the matrix to be investigated, e.g. air to be controlled, gets through a membrane or sintered metal disk into the sensor. Inside that sensor there is the detector element (pellistor) which contains, for example, a thin, spiraled platinum wire surrounded by a ceramic pearl with a catalytically active surface (Fig. 2.5). The oxidizable gas components are catalytically burnt at the heated detector element. The element is additionally heated by the combustion heat generated thereby. This heating results in a resistance alteration which is proportional to the concentration of the combustible gas components. Besides this catalytically active detector element, the sensor also contains a heated compensator element. Both elements represent parts of a Wheatstone bridge, and properties of the medium to be tested, such as temperature, moisture and heat conductivity affect both elements in the same way so that these influences on the measuring signal are (nearly completely) compensated for.

However, the measuring principle is not selective, because all combustible gases and vapors can cause measuring signals. Also non-combustible gases and vapors in higher concentrations may give rise to cross sensitivities (e.g. >6% CO_2 in air). The applicability of sensors can be restricted by catalytic poisons: the catalysts may be poisoned by low concentrations of particular gases of different groups of matter (sulfur compounds, silicone-containing materials, volatile metal compounds, halogenated hydrocarbons). Meanwhile, special pellistors are available for use in presence of the named catalytic poisons.

2.3.1.4
Reactions on Semiconductor Surfaces

Another group of gas sensors is based on reactions on semiconductor surfaces [8]. The electric conductivity of certain semiconductors on the basis of binary and ternary metal oxides (e.g. SnO_2, ZnO, Fe_2O_3, CuO, NiO) depends on the absorption of gases on their solid surface. Gases that can be determined in this way comprise oxidizable materials such as H_2, H_2S, CO and alkanes as well as reducible gases such as Cl_2, O_2 and O_3. The technical use of semiconductor sensors began in the 1970s: Development in this field is by no means completed. Certain details of the reaction mechanisms leading to alterations of properties on the solid surfaces still need further clarification. Important goals of research and development concern the improvement of stability, reproducibility and selectivity of the sensors. Meanwhile, the field of *sensor technology* has gained great importance for industry, ecology, medicine and for analytic research. Many groups of researchers at numerous facilities in universities and industry are working on new and further developments of sensors. For further information one should refer to the literature [8–10].

Semiconductor gas sensors are marked by their simple construction (Fig. 2.6). The measuring method is based on the alteration of the electric conductivity of a semiconductor by surface adsorption of certain gases. This conductivity change causes a change in current which is logarithmically proportional to the analyte concentration. For faultless measurement, the temperature, gas pressure and gas moisture of the semiconductor element have to be kept constant.

These devices used for process analytics are conceived in such a way that they allow recording measurements and induce optical or acoustic signals. In the case of emission measurements, the gas to be analysed is rarefied with nitrogen to a certain extent before the inlet into the analyzer.

Fig. 2.6.
Configuration of a
semiconductor sensor

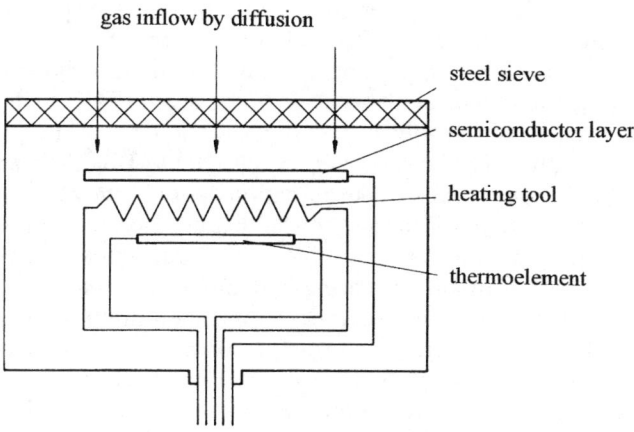

2.3.2
Spectroscopic Methods

2.3.2.1
UV, VIS and IR Spectroscopy

The spectroscopic methods applied in process analytics and subsequently described include absorption spectroscopy in the ultraviolet, visible and infrared regions of wavelength, mass spectrometry and chemiluminescence [6]. Intraatomic or molecular electron transitions are responsible for the absorption of light in the ultraviolet (UV = 200–400 nm) and visible (VIS = 400–800 nm) spectral regions [11], as well as vibrations of molecules or molecular rotations in the infrared region (IR = 0.75 to approximately 100 μm) [12]. By interaction between electromagnetic radiation and valency electrons of a molecule (π and σ electrons) energy is transferred on the molecule. The valency electrons are conveyed into excited, higher energy states by absorption of radiation. Besides electrons, vibrations of molecules and molecular rotations are also excited to a certain extent which leads, especially in the case of UV spectra of solutions, to broad bands instead of single absorption lines. These bands consist of a great number of overlapping single absorptions of different intensity, which correspond to the different (allowed) electron transitions. According to the different possible levels of excitation of electrons, an UV absorption spectrum consists of several absorption regions. Performing the absorption as function of wavelength, the site of the bands on the wavelength scale describes the energy difference between basic state and excited state. The different intensities are caused by different transition probabilities for the single transitions. The intensities are expressed by the molar extinction coefficient ε ($l \cdot mol^{-1} \cdot cm^{-1}$).

UV/VIS spectral photometers are subdivided into single beam and double beam absorption photometers. The single beam photometers are suitable for simple and quantitative routine analyses such as determinations of concentrations at definite wavelengths. A recording of the spectrum is normally not possible. The computer-aided photodiode array apparatuses represent a new development in construction of spectral photometers [13] which enables a fast multi-component analysis. These are single beam devices which record and store the basic line of the solvent before measuring a sample, and then subtract it automatically from the spectrum of the sample. Double beam photometers are suitable for quantitative determinations at definite wavelengths and for recording spectra. A deuterium or xenon lamp is used as radiation source for the UV region, a tungsten lamp for the visible light. Because these light sources are continuum radiators, a prism or grating is required as monochromator. But there are also non-dispersive systems. As detectors, photomultipliers or photo cells are used.

The construction in principle of a non-dispersive UV/VIS double beam absorption photometer is shown in Fig. 2.7: The radiation coming from the light source gets through an interference filter and a radiation splitting simultaneously into the measuring and the reference cuvette. The latter is flown through by an analyte-free medium. The light intensities are determined by two detec-

Fig. 2.7.
Principle of an UV/VIS double beam absorption photometer

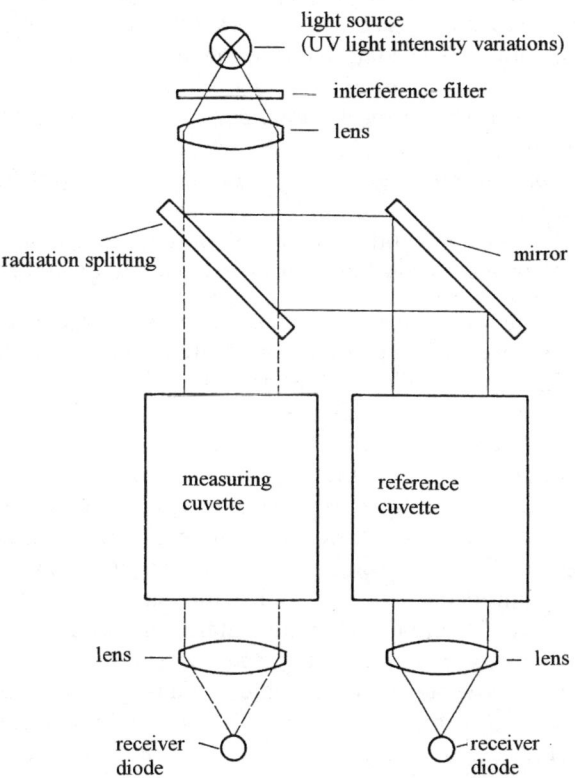

UV double beam principle

light source
(UV light intensity variations)

interference filter

lens

radiation splitting

mirror

measuring cuvette

reference cuvette

lens

lens

receiver diode

receiver diode

tors, where the intensity at the measuring detector depends on the concentration of the analyte, while the accompanying reference signal is concentration-independent. The signal obtained by formation of a quotient corresponds to the transmission (permeability) or, after a corresponding transduction of the experimental value, to the extinction (absorption) of the analyte (compensation principle). The basis for quantitative determinations forms the law of Bougner-Lambert-Beer [8]:

$$- \log \frac{\phi_a(\lambda)}{\phi_e(\lambda)} = \varepsilon_\lambda \cdot c \cdot d = A$$

where

ϕ_e = light intensity of the light radiated into
ϕ_a = light intensity after diminution by absorption
ε = molar extinction coefficient (L mol^{-1} cm^{-1})
c = concentration (mol L^{-1})
d = thickness of the layer (cm)
A = extinction

As the concentration c depends on, among other things, the density of the measuring medium, i.e. on pressure and temperature, the measuring conditions in the case of gases have to be kept constant (constant pressure and constant temperature). Deviations from the linearity A = f(c) may occur by:

- disturbing chemical influences, e.g. interactions of the molecules among each other;
- apparatus influences, e.g. stray light, reflection losses, non-monochromatic radiation;
- fluorescence and Raman emission as unavoidable accompanying radiation. However, this influence is rather low and can therefore usually be neglected.

Other procedures for generating measuring data cannot be dealt with in detail here. One is referred to the literature [2, 11, 14] and the extensive manufacturers' information available (see Sect. 2.5).

According to the various analytes, the obtainable limits of determination differ to a great extent. They vary between 10 and 1000 mg/m³. Some examples are given in Table 2.2. Table 2.3 indicates the versatility of this analytical technique, which is also applied to process analyses of liquid media (see Sect. 3.2.1.1).

The IR spectral region joins with ever decreasing frequencies of the visible region of the electromagnetic radiation. In the region of 2.5–25 μm, there are the most important fundamental, deformation and rotation vibrations of the molecules, which are used for their characterization.

For this analytical task, dispersive and non-dispersive methods are used [12] Dispersive spectrometers which contain as essential components a monochromator with entrance and exit slits and a collimator mirror or a reflection grating in the case of newer instruments as the dispersive element can be used for gas analysis. The advantage of these devices – depending on the wave number range and spectral dispersion – is well-founded by the fact that a great number of dif-

Table 2.2. Limits of determination of UV-absorption photometric measurements (ml/m³) (examples given by suppliers)

Analyte	Spectral line 254 nm	Spectral line 313 nm
Benzene	2.2	
Chlorine	100	4.9
Formaldehyde	99	260
Nickel carbonyl	0.063	0.52
Ozone	0.07	
Phosgene	15	
Mercury	0.001	
Sulfur dioxide	6.3	4.1
Hydrogen sulfide	130	260
Nitrogen dioxide	52	2.1

Table 2.3. Technical application of UV/VIS absorption photometry

Analyte	Matrix	Technical process	Target
Methane	Air	Chemical engineering	Process control
Ethanol	Air	Chemical engineering	Drying process control
Water vapor	Air	Chemical engineering	Drying process control
Carbon monoxide	Burnt gas	Power plant technique	Optimization
Carbon dioxide	Burnt gas	Power plant technique	Optimization
Nitrogen dioxide	Burnt gas	Power plant technique	Control
Nitrogen monoxide	Burnt gas	Power plant technique	Control
Sulfur dioxide	Air	Power plant technique	Control
Hydrogen chloride	Burnt gas	Incineration	Emission control
Carbon dioxide	Digester gas	Biotechnology	Sewage plant control
Dinitrogen monoxide	Air	Medicine	Anesthesia control

ferent substances can be determined with the same apparatus. In many examples of the application, continuous gas analysis can be restricted to the determination of one or a few components [4,14]. In these cases, non-dispersive instruments are in wide-spread use in industry. Non-dispersive means that no radiation-dispersing optical component as in the case of dispersive spectrometers is installed and a relatively broad spectral range is used for measurement. For the selection, filters and selective detectors are incorporated. The following description gives an example for the construction and mode of operation of a non-dispersive IR gas analyzer [15]. As radiation sources underheated filaments (700 °C) are used, radiating a gas-proof closed chamber which is divided by a metal membrane into two halves. Accurate IR spectrometric determination of a gas component in a gas mixture presupposes that its absorption spectrum differs in a suitable manner from those of the remaining components (see Fig. 2.8).

The gas to be analyzed flows through the analysis cuvette which is in the path of rays from an infrared radiator (Fig. 2.9). In a second path of rays is the reference cuvette containing a gas not absorbing in the infrared region. The intensity difference existing at the end of the two paths of rays, which is given by the diminution of the characteristic band of the absorption spectrum of the gas to be analysed, only depends on the concentration of this component in the analysis cuvette. For the measurement of this intensity difference, a receiver is used which consists of two measuring chamber halves separated by a membrane capacitor. The latter are filled with the component to be determined to obtain selectivity of measurement. According to the different IR radiation falling on them, the two halves of the receiver heat up differently, so that a pressure difference exists at the membrane capacitor. A rotating vane interrupts both paths of rays simultaneously, whereby a periodically intermittent measuring effect is produced. The capacity alterations resulting from the change in pressure at the membrane capacitor generate an alternating voltage, which is amplified and recorded.

Fig. 2.8.
Absorption spectra of some gases
in the infrared range

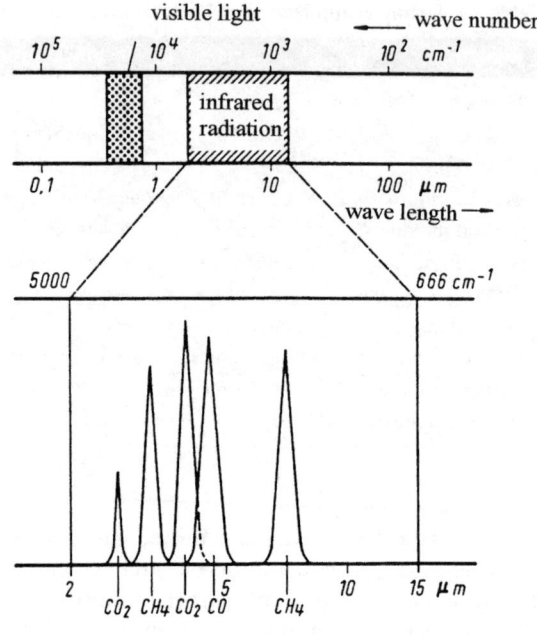

Fig. 2.9.
Scheme of an IR spectrometric gas
analyzer (Hartmann & Braun AG)

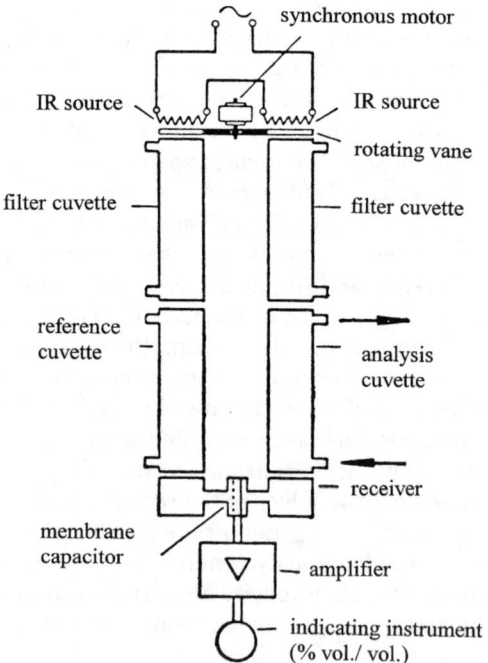

Table 2.4. Examples of application of IR spectrometric gas analysis

Analyte	Analytical measuring range	Analyte	Analytical measuring range
CO	0–0.01 vol. %	NH_3	0–0.1 vol. %
CO_2	0–0.005 vol. %	H_2O	0–0.25 g/m^3
CH_4	0–0.02 vol. %	CH_3OH	0–1.0 g/m^3
C_3H_8	0–0.02 vol. %	C_2H_5OH	0–2.0 g/m^3
C_2H_2	0–0.05 vol. %	C_6H_6	0–2.0 g/m^3

The measuring range can be adapted to the analytical problem by the election of suitable absorption bands or the length of the cuvettes. The selectivity of the receiver can be increased by using additional filter cuvettes. Table 2.4 shows some analytical examples of IR spectrometric gas analysis. The technical application serves, for example, for the optimization of industrial boiler plants [16], the analysis of top gas (see Sect. 6.2.2) burnt gas and other industrial waste gases, automotive exhaust gas, the investigation of industrial emissions and immissions, the control of pure and synthesis gases as well as of air liquefaction plants [17], the determination of air pollutants, the close loop control of the cement production process [18] as well as for the CO_2 control of fermentation processes in the food industry [19]. Calibration of these systems is done by means of commercial (certified) testing gases. The requirements for the limits of determination may be very different. For control of room air and exhaust air plants, limits of detection for NO, NO_2 and SO_2 of approximately 200 $\mu l/m^3$ must be reached. This demand can only be fulfilled by application of Fourier transform infrared spectroscopy (FT-IR) [20].

2.3.2.2
Mass Spectrometry

On-line analysis of complex gas mixtures when controlling and optimizing processes or environment technology sets ever higher standards for the analytics, especially regarding the precision of gas analytic determinations in high concentration ranges. This has led to, among other things, the introduction of mass spectrometry into gas analytics instead of IR absorption spectroscopy [21].

Figure 2.10 shows the schematic construction of a mass spectrometer for the determination of CO_2, CO, N_2 and H_2. It consists of the base units for sample supply, ion generation, separation of the ion species and measurement. The gas to be analyzed is first supplied to the ion source in high vacuum. As the ionization of the gas molecules must take place at a pressure of 10^{-4} to 10^{-6} hPa, the gas pressure of the sample has to be reduced to the working pressure in the ion source. Here the gas molecules cross an electron beam and are partially ionized (electron collision) [22]. The non-ionized gas molecules are pumped out.

Ions generated this way, with an initial energy as uniform as possible, are, by electric fields, brought out of the area of the ion source, accelerated and guided onto the inlet slit of the mass spectrometer. Here they get into the field of the deflection magnet and a capacitor and are deflected with different strength

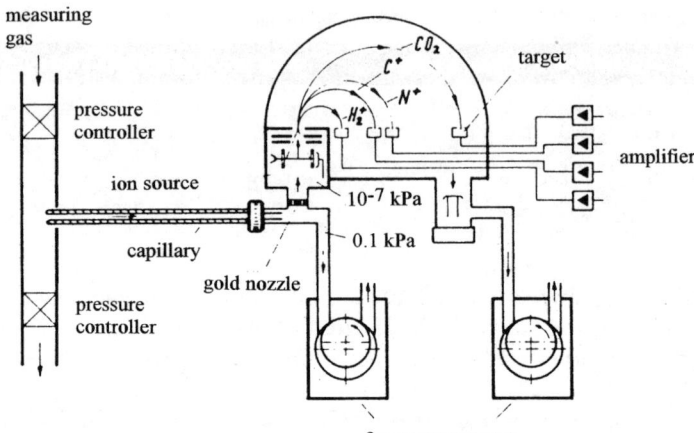

Fig. 2.10. Example for the configuration of a mass spectrometer for gas analysis

according to their mass (more exactly: m/e^+ ratio). The deflection is stronger the smaller the mass and the speed of the particles. By focusing facilities, the separation of ions of the same mass is performed in such a way that their measurement can be done by an electric detector. Faraday receivers are behind the exit slits. The current flowing to the receivers for neutralization of the ions corresponds to the ion current of the ion species concerned and is proportional to their concentration.

For continuous automatic process gas analysis, computer-controlled mass spectrometer systems are used nowadays [23], these involving automatic calibration with certified testing gases.

In the case of complex composed process gases, which may contain hundreds of (organic) compounds, conventional mass spectrometry is not applicable to direct and continuous trace analysis in default of adequate sensitivity and selectivity. Here, resonance-enhanced multi-photon ionization (REMPI) enables a selective and sensitive ionization of compounds in complex substance mixtures [24]. Using this ionization method in combination with time-of-flight mass spectrometry produces an analytical method (REMPI-TOFMS) with high sensitivity and selectivity as well as high measuring speed for many applications in process and environmental analytics. Further development of this new method should be regarded with special interest.

2.3.2.3
Chemiluminescence

In chemical reactions of some gases, energy in the form of light is set free producing a characteristic luminous phenomenon called chemiluminescence [6, 25]. Its intensity is, under defined conditions (constant pressure in the reaction chamber, constant measuring gas flow), an unequivocal measure for the amount of such an analyte in a mixture with other gases which do not show this phenomenon. This analytical principle is only applied in special cases, but is, because of its analytical characteristics, of particular interest.

The phenomenon of chemiluminescence is used, for example, for the continuous and recorded determination of nitrogen oxides [26]. In the case of NO determination, chemiluminescence occurs when oxidizing nitrogen monoxide with ozone:

$$NO+O_3 \rightarrow NO_2+O_2+h_v$$

When the nitrogen dioxide molecules excited are transient into the basic state, the excitation energy is set free in form of light.

Ozone is generated by electric discharge in an air or oxygen stream according to the amount of nitrogen monoxide.

2.3.3
Electrometric Methods

In process analytics, electrometric methods have been widely applied [2, 6]. Among others, *electrolytic conductivity* has been introduced as the quantity to be measured. The specific electrolytic conductivity of aqueous solutions is caused by dissociation of those substances dissolved in the water and therefore it is concentration and temperature dependent. This measuring principle is mainly used for trace analysis, e.g. for hydrogen sulfide, sulfur dioxide, ammonia and water vapour down to amounts of 10^{-4} vol. %.

The principle is based on the continuous and specific absorption of the analyte in a suitable liquid, the conductivity change of which is a measure for its concentration. Figure 2.11 shows such an analyzer: after completion of the absorption in the reaction section, the flow of liquid is divided into a branch free of gases and a gas-containing branch, which leads the non-absorbable parts of gases into open air. The branch free of gases contains, for the measurement of conductivity, a pair of electrodes as measuring section. The electric conductivity of the absorption liquid in the measuring section is compared with the electric conductivity before reaction in a comparison section by means of a Wheatstone bridge. The difference of the conductivities between the measuring and comparison sections is a measure for the concentration of the component to be determined.

Beside this method, there are even more procedures based on the conductivity principle (oxygen concentration chains; current measurement by galvanic elements), which should only be mentioned at this place.

Furthermore, *coulometric methods* are applied to trace analysis of gases, which possess certain advantages in the case of many measuring tasks. The following example may serve for explanation. The continuous generation of the required reagent on the spot by electrolysis under permanent supply and measurement of the current needed for this is much simpler than the continuous addition of a reagent. Such coulometric gas analyzers are used, for example, for the determination of hydrogen sulfide, sulfur dioxide and mercaptanes in air in the µg/g range.

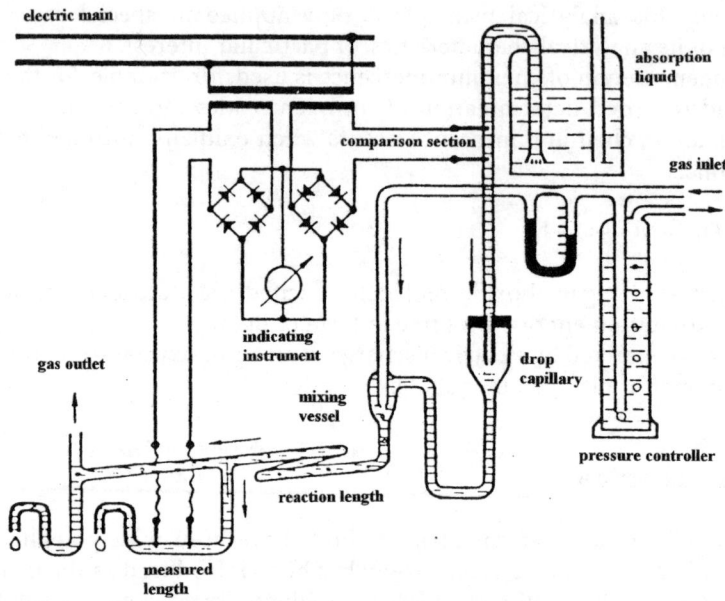

electric main

absorption
liquid

comparison section

gas inlet

indicating
instrument

gas outlet

drop
capillary

mixing
vessel

pressure controller

reaction length

measured
length

Fig. 2.11. Scheme of a conductivity analyzer (Hartmann & Braun AG)

Figure 2.12 clarifies the principle: in the anode region a certain quantity of free bromine, which stimulates an indicator electrode, is generated by electrolysis of a potassium bromide-containing absorption solution. Its potential regulates the electrolytic current so that the free bromine, which is consumed when introducing, for example, the sulfur dioxide or hydrogen sulfide-containing process gas into the anode region, is reproduced to a corresponding extent. The quantity of current of the electrolytic current needed is a measure for the existing bromine-consuming gas components. This method can be performed more specifically by prior separation of disturbing gases with the help of absorption solutions (e.g. cadmium acetate for hydrogen sulfide).

Besides coulometric, there are potentiometric and amperometric methods of measurement. In case of amperometric sensors, the current is measured which flows through a three-electrode measuring cell at constant working electrode voltage. The current intensity proportional to the concentration of the analyte depends on the electrochemical mass reaction (oxidation/reduction of the analyte) that takes place per time unit at the electrode surface. The classic example for an amperometric sensor is the oxygen sensor according to Clark [27] in which a membrane permeable to gas enables the diffusion of oxygen to the electrode surface. As the membrane is impermeable to liquids and dispersed solids, the oxygen dissolved in liquids can be determined independent of the partial pressure. The method is extremely sensitive, for a current of 1 μA corresponds to a mass reaction of only approximately 10^{-10} mol/s. However, the disadvantage of

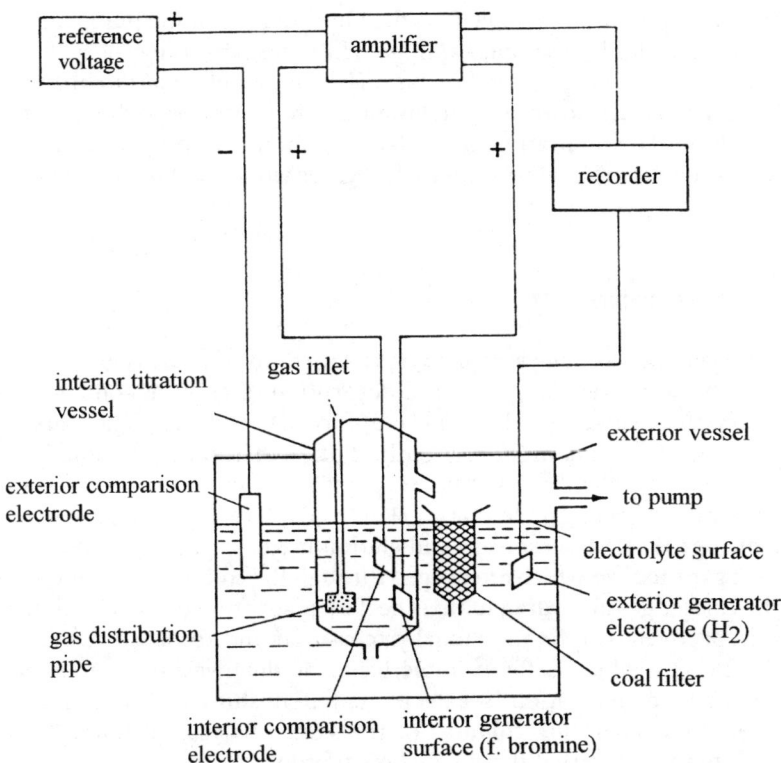

Fig. 2.12. Scheme of a coulometric gas analyzer

this sensor is that the signal depends on the stirring, as the oxygen is permanently consumed and the post-diffusion through the membrane depends on convection in the analyte solution. Attempts are being made to eliminate this disadvantage by means of a number of measures.

The selectivity of amperometric gas sensors can be systematically influenced by changing the working electrode potential and by skillful selection of the membrane material and of the electrolyte.

Finally, the principle and the application of the *flame ionization detector* should be mentioned. That method, applicable to the determination of organic components, is based on the fact that when organic substances are burnt in a hydrogen flame carbon ions are produced by thermal ionization, which generate an ion current in an electric field. This ion current (approximately 10^{-8} to 10^{-12} A) leads to a voltage drop at a highly resistive resistance which, after amplification, is recorded as a signal. The strength of the ion current depends on the number of existing ions, therefore directly on the mass of hydrocarbons and also on the parameters of the hydrogen flame.

In this method the gas sample is sucked in by a membrane pump and is supplied to the detector by a sample capillary. The measuring range of such a device can cover 10^{-2} to 10^{-5} mg/kg total carbon. The gas sampling has to be done with heated piping in order to avoid condensation of higher hydrocarbons on the way to the analyzer. The calibration of the detector and the testing of the linearity of the measurements are performed with test gases which contain graded amounts of the analyte.

2.3.4
Gas Chromatography

Chromatographic gas analysis, an approved method of process analytics [6], is based on the selective adsorption and desorption of gases at solid adsorbents (stationary phase) and their different solubility in liquids [28]. There are several modes of operation [29]. In the case of the elution method predominating today in gas chromatography, the gas sample is passed through a chromatographic column by a carrier gas (mobile phase) (Fig. 2.13). Here, the carrier gas (helium, argon, nitrogen) serves as solvent and elution medium. The individual components of a gas mixture, which are purged through the column more or less rapidly by the carrier gas according to their heat of adsorption or distribution coefficients, respectively, can be individually measured and recorded at the column exit by a detector which is, for example, based on the measurement of heat conductivity or flame ionization (see Sect. 2.3.3). Reflushing and reversing the gas flow on several separating columns of the same analyzer (column reversing) allows adaptation to very different analytical problems.

The expenditure to automate a discontinuous analysis procedure such as gas chromatography for process analytical purposes is considerable. The complex construction of a process chromatograph can entail that the operation of such devices leads under certain working conditions to more disturbances than with analyzers that are based on less costly or less complicated principles.

For the control of distillation and rectification processes in the chemical and petroleum industries, process chromatographs are widely applied. Gas chromatography not only serves for the analysis of process materials in the framework of process control and optimization, but also for air control in production plants.

Pollutants can get unnoticed into the air by accidental disturbances such as leaks in piping, pumps or reaction vessels, and can possibly endanger the employees working at these installations. Therefore, it is necessary to perform either occasional or regular control measurements or even continuous control of the air as a special part of the analytical activities according to the potential danger caused by the technical process concerned. Within the gas chromatographic monitors developed for this purpose (known are, for example, monitors for acrylonitrile, diethyl sulphate, dimethyl sulfate, epichlorohydrin, nitrobenzene) the steps of the analytical process such as "sampling, analysis, evaluation of the measured results and recording or sounding an alarm" are fully integrated. Recalibration of these devices takes place automatically at pre-determined intervals with commercial test gases.

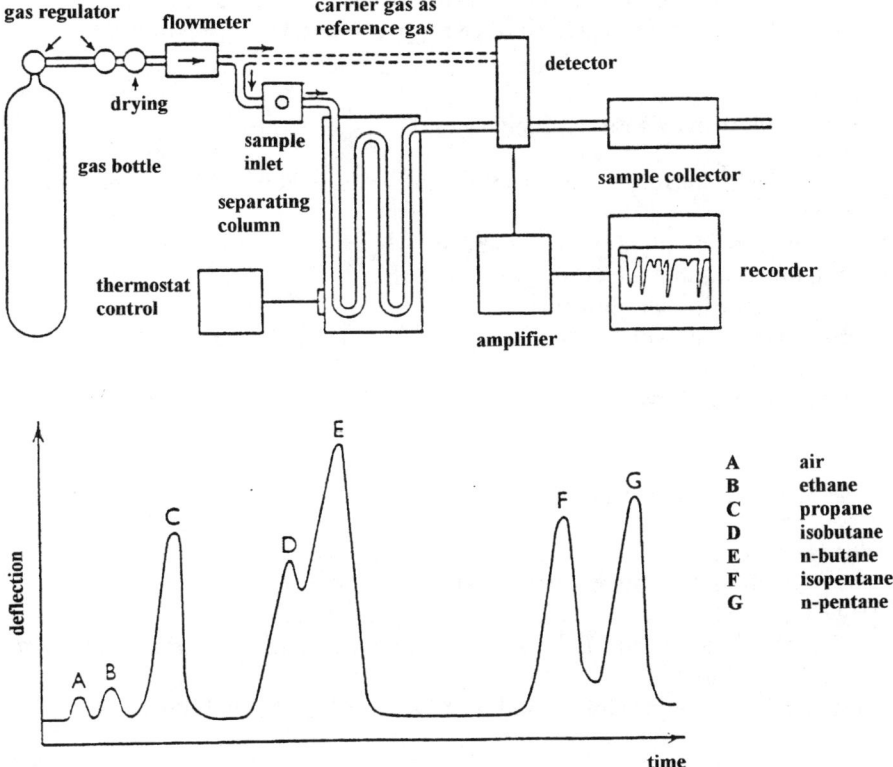

Fig. 2.13. Principle of a gas chromatograph and an example of a gas chromatogram

The newest trend in this field is characterized by the development of miniaturized chromatographic systems which can be used as process analytical sensors (see Sect. 3.2.1.2).

2.4
Suppliers for Gas Analyzers and Gas Sensors

2.4.1
Paramagnetic Gas Analyzers

Beckman Instruments GmbH, Frankfurter Ring 115, D-80807 München
Hartmann & Braun AG, Postfach 900507, D-60445 Frankfurt/Main
Honeywell AG, Dornierstrasse 4, D-82178 Puchheim
Leeds & Northrup GmbH, Kimpler Strasse 288, D-47807 Krefeld
Maihak AG, Postfach 601709, D-22292 Hamburg

SERVOMEX-Gasanalysentechnik, Harkortstrasse 29, D-40880 Ratingen
Siemens AG, Meß- und Regeltechnik, Kruppstrasse 2, D-45128 Essen

2.4.2
Heat Conductivity Measuring Equipment

ADOS – Technik GmbH für Verfahrenstechnik, Eckener Strasse 30, D-59075 Hamm
Beckman Instruments GmbH (see above)
Hartmann & Braun AG (see above)
Leeds & Northrup (see above)
Maihak AG (see above)
Servomex (see above)
Siemens AG (see above)

2.4.3
Gas Sensors Based on Heat of Reaction

Ados (see above)
Auer – Gesellschaft, Thiemannstrasse 1, D-12059 Berlin
Drägerwerk AG, Moislinger Allee 53–55, D-23542 Lübeck
General Monitors Ireland Ltd., Queens Ave., Hurdsfield Industrial Estate, Macclesfield, Cheshire GB-SK 10 2BN
Zellweger Eco-Systeme GmbH, Sollner Strasse 65 b, D-81479 München

2.4.4
Semiconductor Sensors

Auer – Gesellschaft (see above)
Bieler & Lang GmbH, Ludwigsburger Strasse 14, D-28215 Bremen
Drägerwerk AG (see above)
GfG – Gesellschaft für Gerätebau mbH, Klönnestrasse 99, D-44123 Dortmund
Siegrist-Photometer AG, Hofurlistrasse 1, CH-6373 Ennetbürgen

2.4.5
UV-/VIS-/NIR- Process Photometers

Beckman Instruments GmbH (see above)
Bran & Lübbe GmbH, Werkstrasse4, D-22844 Norderstedt
Colora Meßtechnik GmbH, Barbarossastrasse3, D-73547 Lorch
Hartmann & Braun AG (see above)
Jasco Deutschland GmbH, Robert-Bosch-Strasse 11, D-64823 Groß-Umstadt
Kontron Instruments GmbH, Werner-von-Siemens-Strasse 1, D-85375 Neufahrn
Optometrics Ltd., Unit C6, Cross Green Garth, GB-Leeds, West Yorkshire LS9 0SF
Perkin-Elmer Bodenseewerk GmbH, Postfach 101761, D-88647 Überlingen
Philips Industrial Electronics Deutschland GmbH, Miramstrasse87, D-34123 Kassel

Wilhelm Pier GmbH, Voltastrasse7, D-65795 Hattersheim
Shimadzu Europe GmbH, Albert-Hahn-Strasse 6–10, D-47269 Duisburg
Sigrist-Photometer AG (see above)
Unicam Analytische Systeme GmbH, Korbacher Strasse 75 –77, D-34132 Kassel
Varian GmbH, Alsfelder Strasse 6, D-64289 Darmstadt
Carl Zeiss Jena GmbH, Tatzendpromenade 1 a, D-07740 Jena

2.4.6
IR Process Photometers

Beckman Instruments GmbH (see above)
Maihak AG (see above)
Auer-Gesellschaft (see above)
Siemens AG (see above)

2.4.7
Mass Spectrometric Gas Analyzers

ATOMIKA Instruments GmbH, Bruckmannring 40, D-85764 Oberschleißheim
Bruker Analytische Meßtechnik GmbH, Silberstreifen, D-76287 Rheinstetten
Finnigan MAT GmbH, Barkhausenstrasse2, D-28197 Bremen
Perkin-Elmer Bodenseewerk GmbH (see above)
Shimadzu Europe GmbH (see above)
Varian GmbH (see above)

2.4.8
Chemiluminescence Sensors

Beckman Instruments GmbH (see above)
Bendix Deutschland GmbH, Im Rübenkamp 11, D-38162 Cremlingen
Kontron Instruments GmbH (see above)

2.4.9
Electrometric Measuring Systems

Beckman Instruments GmbH (see above)
Deutsche Metrohm GmbH & Co, In den Birken 3, D-70794 Filderstadt
Hartmann & Braun AG (see above)
Ingold Meßtechnik GmbH, Siemensstrasse 9, D-61449 Steinbach
Knick Elektronische Meßgeräte GmbH & Co, Beuckestrasse 22, D-14163 Berlin
Philips GmbH (see above)
Schott-Geräte GmbH, Postfach 1130, D-65701 Hofheim
Siemens AG (see above)
Ströhlein GmbH & Co, Girmeskreuzstrasse 55, D-41564 Kaarst
Wissenschaftlich-Technische Werkstätten GmbH (WTW), Dr.-Karl-Slevogt-Strasse 1, D-82362 Weilheim
H. Woesthoff GmbH, Max-Grewe-Strasse 30, D-44727 Bochum

2.4.10
Process Chromatographs
Beckman Instruments GmbH (see above)
Bendix (see above)
Dani Strumentazione Analitica spa / ESWE Analysentechnik GmbH, Zwinger-strasse4a, D-74889 Sinsheim
Foxboro Deutschland GmbH, Heerdter Lohweg 53–55, D-40549 Düsseldorf
Ratfisch Analysensysteme GmbH, Tel.: 08121–82081
Servomex (see above)
Siemens AG, Meß- und Regeltechnik, Kruppstrasse2, D-45128 Essen

2.4.11
Remarks

The preceding compilation cannot make any claim to completeness. For further information reference should be made to the literature and the market surveys published in "Nachrichten für Chemie, Technik und Laboratorium", the news of the Gesellschaft Deutscher Chemiker (GDCh) (Society of the German Chemists).

Further information is given in Sects. 3.4, 4.5 and 6.9.

References

1. Gromadzki D (1989) Steel & Metals Magazine 27(3):171
2. Melzer W, Jaenicke D (1980) In: Ullmann's Enzyklopädie der technischen Chemie, 4. Aufl., Chemie, Weinheim, Bd 5:891
3. Dosch D (1907) Z Chem App kde 2:473
4. van Damme S, Slemeyer A, Wendt K (1987) Techn. Messen 54H(11):416
5. DIN 51 853: Testing of fuel gases, protective and exhaust gases; sampling
6. Oesterle G (1995) Prozeßanalytik –Grundlagen und Praxis. Oldenbourg, Munich
7. Lagois J (1992) CLB Chem Lab Biotechn 43:428
8. Cammann K, Ross B, Hasse W, Dumschat C, Katerkamp A, Reinbold J, Steinhage G, Gründig B, Renneberg R, Buschmann N (1994) In: Ullmann's Encyclopedia of Industrial Chemistry, VCH Publishers, Weinheim, vol B6, p 121
9. Tränkler H-R, Obermeier E (eds) Sensortechnik –Handbuch für Praxis und Wissenschaft, Springer, Berlin Heidelberg New York
10. Göpel W, Hesse J, Zemel JN (eds) (1998) Sensors, Wiley-VCH, Weinheim
11. Gottwald W, Heinrich KH (1998) UV/VIS-Spektroskopie für Anwender. Wiley-VCH, Weinheim
12. Günzler H, Heise HM (1996) IR-Spektroskopie. 3. Aufl., Wiley-VCH, Weinheim
13. Weismüller JA (1991) Kontrolle H Juni, p 60
14. Gauglitz G (1994) Nachr Chem Tech Lab 42, 11:M 1
15. Staab J (1994) Industrielle Gasanalyse. Oldenbourg, Munich
16. Fabinski W, Eckmann F (1987) VBG Kraftwerkstechn 67:143
17. Nather E, Schorpp K (1982) Siemens-Energietechn 4:141
18. Richter J, Hartmann, Braun AG, Einzelber. 02 PY 3604
19. Richter J, Hartmann, Braun AG, Einzelber. 02 PY 3603
20. Williams RR (1995) Fourier transform spectroscopy. VCH Verlagsgesellschaft, Weinheim
21. Schuy KD, Reinhold B (1972) Stahl u. Eisen 92:1278
22. Budzikiewicz H (1992) Massenspektrometrie – Eine Einführung, 3. Aufl., Wiley – VCH, Weinheim
23. Rauch W, Tegtmeyer U, Schlögl R (1994) GIT Fachz Lab, H 2:93

24. Zimmermann R, Heger HJ, Kettrup A, Boesl U (1997) Rapid Communic Mass Spectrom 11:1095
25. Campbell AK (1988) Chemiluminescence, VCH Verlagsgesellschaft, Weinheim
26. Fortijn A, Sabadell A, Ronco RJ (1970) Anal Chem 42:575
27. Clark LC (1958) US Pat 2,913,386
28. Gottwald W (1995) GC für Anwender. Wiley-VCH, Weinheim
29. Oster H (1973) Prozeßchromatographie, Methoden der Analytik in der Chemie, vol 15. Akadem. Verlagsanstalt, Frankfurt am Main

3 Process Analytics of Liquid Phases

3.1 Sampling of Liquid Phases

3.1.1 Aqueous and Organic Materials

Sampling of aqueous systems is performed in a similar (simple) way as if genuine solutions are involved. In the case of emulsions problems are to be expected or appropriate measures are to be taken to ensure that a representative sample is obtained. Liquids containing solids and multiphase systems of immiscible substances are also problematic [1].

Figure 3.1 illustrates the technique of sampling of liquids by means of a bypass and branch pipe-line [2]. The product is taken off via a distributor tube of the pipe-line or the bin and it is preferably returned to the process by means of a bypass line. For the conveying of the product through the bypass line, pressure differences or a separate conveying pump can be used. Thus, frequently, for example, pressure and suction sides of process-end conveying pumps, and also flow resistances in the product line itself, are made use of.

A fraction of the sample stream flowing in the bypass line is taken off by a probe out of an adapter in the pipe-line and fed to the analyzer by a branch line which has to be as short as possible. In order to be able to carry out work on these connecting pieces, it should be possible to shut off the bypass line with two valves. The probe is followed by a ball or regulating valve and, if necessary, an adjusting valve or pressure reducing valve and, frequently, also an overflow for pressure regulation. A conveying pump as well as a fine filter at the inlet of the equipment are also possibly necessary.

For the separation of solid portions a filter at the inlet to the sampling probe is particularly suited. By aligning the filter surface parallel to the flow direction, in certain circumstances, self-cleaning of the filter surface in the bypass stream can be achieved.

In the case of degassing liquids the overflow bin can be designed as a degasser. The liquids released during sample preparation must either be returned to the process by a discharge line (collection bin with conveying pump) or passed into the canal or the waste water cleaning plant.

Depending on the nature of the measuring fluid, heating or cooling of the sampling system can be inevitable. By contrast to gases, in this case far greater heating or cooling capacities are necessary.

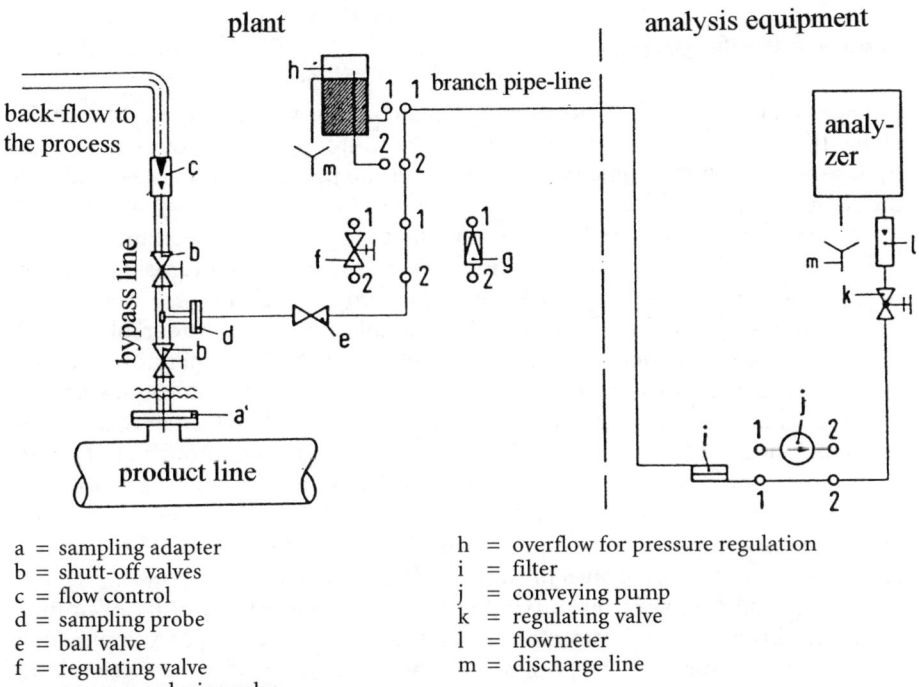

a = sampling adapter
b = shutt-off valves
c = flow control
d = sampling probe
e = ball valve
f = regulating valve
g = pressure reducing valve

h = overflow for pressure regulation
i = filter
j = conveying pump
k = regulating valve
l = flowmeter
m = discharge line

Fig. 3.1. Sampling of liquid flows

If the aim is to assess liquid flows and their constituents – which is of particular importance in the field of waste water – quantity-proportional sampling must be ensured. Time-proportional sampling is often out of the question because of fluctuating mass flows.

The problems occurring during sampling and subsequent sample preparation can be summarized in the following way:

1. selection of the phase characteristic for the process and representative for the process condition;
2. taking-off of a partial sample (or a part stream) with reference to the process conditions (e.g. place of sampling, dead time);
3. avoidance of changes in the composition of the sample;
4. adjustment of the sampling conditions to the analytical method.

The automation of sampling which, in certain circumstances, can involve considerable technical difficulties offers decisive advantages: avoidance of systematic errors, reduction of random errors and time coupling to process events. In each individual case, the questions of the admissible dead time, the availability of automatic systems and the required control functions must be answered separately.

3.1.2
Melts of Metallic Materials

The liquids naturally also include melts which, however, have their own problems: stratification resulting from temperature differences, development of multiphase systems, heterogeneity resulting from the presence of metallic and oxidic components.

The sampling of pure molten metals with homogeneous temperature distribution is simple, as, for example, the zinc baths in the hot dip galvanising process. In this case, sampling takes place by means of simple steel ladles.

The sampling of molten steel baths, for example, involves considerably greater expenditure and a variety of problems. In this case high temperatures (1600–1700 °C) and the existence of a multicomponent system are to be taken into account. Thereby, great process-analytical importance is attached to the representativeness of the sample. Today, the sampling is carried out mainly with "one-way probes", which, with the help of a mobile device, are dipped reproducibly into the melt. Then the sample moulds consisting of two halves are filled by the ferrostatic pressure. The important details will be covered when considering process analytics of steel production (see Sect. 6.3.2.1).

As a special industrial example, the aluminum industry can be examined since here both a fusion electrolyte and a molten metal are to be sampled in a manner compatible with the process [3]. The particular bath composition is an important state variable which must be known for optimum operation. The samples taken from the molten bath can show axial and radial segregation. The solidification morphology and the grain structure must be known as they influence the subsequent spectrometric analysis.

3.2
Analysis of Liquid Phases

3.2.1
Aqueous and Organic Materials

3.2.1.1
On-Line Concepts

As analytical principles, in particular for the analysis of aqueous samples in the field of on-line analysis titrimetry [4], various electrochemical methods [5], photometry [6], ICP spectrometry [7] and chromatographic methods for organic questions [8] as well as X-ray fluorescence and Raman spectroscopy [9] have been introduced.

The advantages for on-line control of process solutions (e.g. pickling and treatment baths, electrolytes) with the help of the analytical principles mentioned can be stated as being:

- current information on the process stage;
- quick and timely detection of disturbances;

- complete evidence of the product structure within the scope of quality assurance;
- possibility of optimizing the process sequence;
- coupling of on-line analyzer and automatic control systems;
- saving of raw materials by stricter quality control and better adherence to product specification;
- energy saving (for example in the case of technical electrolytes);
- less environmental pollution by means of reduction of pollutants;
- higher safety by measuring in the endangered environment without contact with the product measured.

As examples of the wide range of applications of such on-line methods the control of phosphating and passivating baths for the pre-treatment of car bodies and car body parts [10] can be examined. Here, the manual bath control has been replaced in many cases by process-analytical on-line monitoring in the guise of quality assurance [11] (Fig. 3.2).

The development of flow injection analysis (FIA) is particularly suited to the trend in process analytics towards the simplest and quickest analytical methods [12]. In this case a carrier flow, often an electrolyte solution, serves at the same time as a transport and work operator. Reagents for chemical reactions can be fed into the flow, the detector having to respond specifically to the components to be determined. The FIA can be automated up to data output and allows a high sample throughput [13]. In principle, any indication system can be used for detection, such as coupling with electrochemical or spectral analytical detectors (see Sect. 5.2).

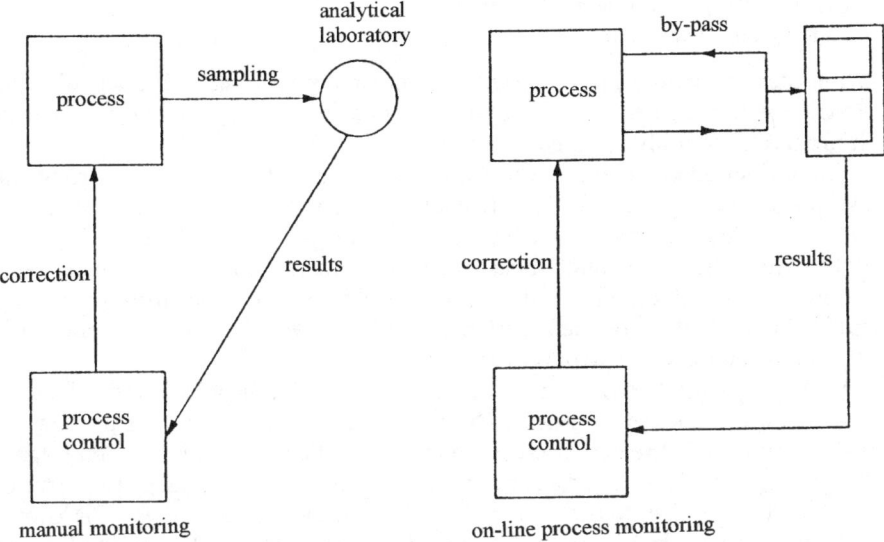

Fig. 3.2. Comparison of manual and on-line process control

Table 3.1. Examples of application of process titrators for chemical process control

Process	Analyte
Peroxide production	$H_2SO_4, H_2O_2, KHSO_5$
Sweetening agent production	CN^-
Caustification	NaOH, alcoholates, phenoles
Polyphosphate production	NaH_2PO_4, Na_2HPO_4
H_2SO_4 production	nitrosylsulphuric acid
PVC production	H_2O
Cyanamide production	cyanamide
Fertilizer production	H_3PO_4, HNO_3
Chromosulphuric acid production	Cr(III), Cr(VI)
Caprolactam production	NH_4^+, NH_3
Pickling of metals	Cl^-, Fe^{2+}
Phosphatization	$PO4^{3-}, NO^{2-}, Zn^{2+}$
Steam generation	SiO_2
Cooling water control	OH^-

"*Process titrators*" [14] are used today in many industrial fields of process control. The range of applications which have been achieved in the field of process titration is extraordinarily large (Table 3.1), due to the large number of possible types of reaction and indication methods. In principle, titration reactions can be divided into five main groups:

- redox reactions: cerimetry, manganometry, dichromatometry, iodometry, diazotizations;
- complexometry;
- neutralization reactions, alkalimetry, acidimetry;
- precipitation reactions; argentometry;
- Karl-Fischer titrations.

In the case of turning point recognition, normally the measuring signal of an indicator electrode is received. For that purpose potentiometric, voltametric or photometric methods are used.

The measured value acceptance of the signal of the indicator electrode can take place time-controlled or drift-controlled. In the case of time-controlled measured value acceptance, this happens after a rigidly prescribed waiting time. In the case of drift-controlled measured value transmission, the measured value is only accepted if the electrode signal is stable within a permitted variation width. The evaluation of the titration curve is carried out mainly by end point titration or digital turning point titration.

With *end-point titration* the potential of the turning point (desired value) must be known and reproducible. The process titrator compares the preselected desired value with the actual measured value in the sample and, depending on the control deviation and as a function of the parameter setting, passes large or small metering pulses to the motor piston burette. As the desired value is known and does not have to be sensitively titrated over the whole pH/potential range, the titrating agent can be very quickly added up to the proximity of the end

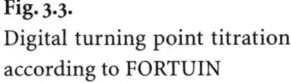

Fig. 3.3.
Digital turning point titration
according to FORTUIN

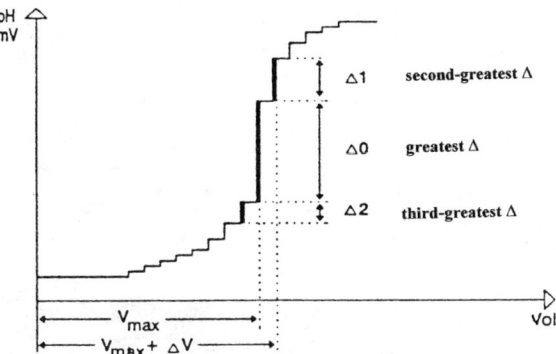

point. The titration time is therefore very short. When the desired value is reached within a pre-selected permitted tolerance range, the process titrator switches off the titration process after a defined switch-off lag. A special case of end-point titration is the Karl-Fischer titration used for water determination, in which the end point is indicated voltametrically on a double platinum electrode.

The dosing of the titration agent is carried out incrementally in the case of *digital titration*, i.e. in preselectable, constant volume steps. After each volume step the measurement of the potential change of the electrode system takes place. The measured value is accepted drift-controlled or time-controlled.

According to Fortuin the equivalence point can be determined on the basis of the three largest potential differences in the turning point area (Fig. 3.3). Evaluation is carried out by means of an interpolation method. The total volume of reagent added before the greatest potential jump $\Delta0$ is designated as V_{max}. The end point of titration is then between V_{max} and $V_{max}+\Delta V$. The various interpolation methods calculate the final volume V_E from $V_E = V_{max} + \rho\Delta V$. The parameter ρ indicates the position of the final volume V_E within the volume step. By the arithmetical determination of the equivalence point over the potential differences in the turning point area independence of the calibration of the measuring chain [15] is achieved.

Figure 3.4 shows a schematic overview of the sequence of an individual determination in a process titrator. The sample flows with opened sample valve through the overflow pipette until the current sample is present in the macro-sample measuring system. Then the sample valve is closed and the sample measured off in a defined way in the overflow pipette is transferred to the titration vessel. The sample submitted is mixed with the required auxiliary reagents and the parameter to be determined is analyzed with the help of a suitable electrode and the appropriate titration agent. After the evaluation of the turning point the percentage is calculated and it is given out by an interface for logging and by an analogue output to the control unit. For turning point identification during titration, e.g. of the free acids and of total acid, of nitrite and of zinc in a bath, digital titration is used.

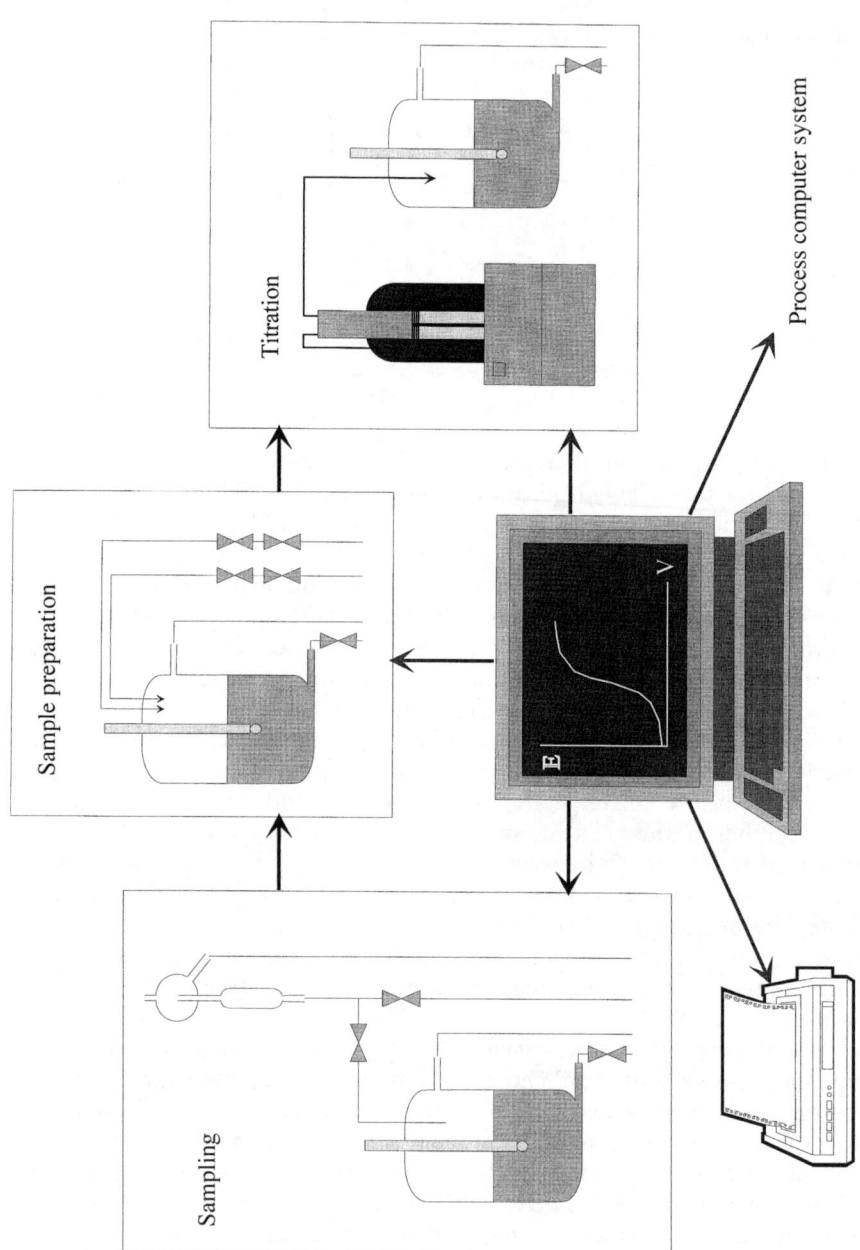

Fig. 3-4. Scheme of the analytical sequences within a process titrator

As further examples, the monitoring of pickling and electrolytic processes can be mentioned for which, besides the analysis of the pickling solutions, the control of emissions is necessary [16]. Table 3.2 gives, for example, an impression of the titrimetric process monitoring in the circuit board industry [17]. The possible operational applications in the steel industry include the determination of Zn^{2+} and H_2SO_4 contents in electrolytic zinc baths, of the Cr^{6+} content in chromium plating baths as well as Fe^{2+} and acid contents (HNO_3, HCl, H_2SO_4, HF) in pickling baths.

In the case of process control in the field of galvanizing and electroforming (galvano-technical component production), in addition to (electrometric) analysis of inorganic components of electrolytes (Table 3.3) there are the determination of additives, moistening agents and organic decomposition products to be performed. For the last mentioned task *chromatographic methods* such as high performance liquid chromatography (HPLC) and gel chromatography [8] are used.

On-line analysis using *process photometers* is employed essentially for two tasks:

– photometric indication of a titration, e.g. in the case of complexometric titrations, in which the end point is indicated with colored indicators, or in the determination of acid values in oils, in which normal glass electrodes only have short service lives;
– photometric direct measurements of individual substances as for the determination of low concentrations of ammonium, nitrate, nitrite, phosphate, phenol, silicate etc.

Table 3.2. On-line control of electrolytes for galvanic coppering

Parameter	Analytical range	Analytical method	Standard solution
Cu	15–25 g/l	Iodometry Complexometry	$c(S_2O_3^{2-})=0.1$ mol/l $c(EDTA)=0.1$ mol/l
H_2SO_4	120–220 g/l	Acidimetry	$c(NaOH)=0.1$ mol/l
NaCl	40–100 mg/l	Argentometry	$c(AgNO_3)=0.01$ mol/l

Table 3.3. Application of potentiometric methods in the field of galvanizing (examples)

Analyte	Process medium
Cu, Au	Galvanic baths
Al, Cl^-, Cr(VI), oxalic acid, H_2SO_4, SO_4^{2-}	Anodizing baths
Pb, H^+, Sn	Lead baths
Cd, CO_3^{2+}, CN^- NaOH	Cadmium baths
NH_4^+, CO_3^{2+}, CN^-, Cu, NaOH, SO_3^{2-}, Zn, Sn	Brass and bronzing baths
F^-, hexafluorosilicate	Chromium baths
Ni, Cl^-, boric acid, tetrafluoroboric acid	Nickel baths
acetate, carbonate, Cl^- fluoroboric acid, H^+, OH^-, NO_3^-, Sn	Tin baths

The possibility of using diode arrays as detectors and of fibre optic sensors [18, 19] represents an important step forward in the use of photometry in process analytics. In that way the classic arrangement of light source-cuvette-detector can be dispensed with and light can be guided directly to the place of spectrophotometric measurement and from there to the detector (Fig. 3.5). Furthermore, probes according to the ATR principle (attenuated total reflection) [20] as well as probes for fluorescence measurement and for the measurement of diffuse reflections are of technical interest. In particular the highly resistant and seal-free probes made of ceramics permit their use in aggressive fluids [18].

A process photometer with a light guide coupling consists of a chopper motor, curve filter, detector and evaluation unit. The measuring of the colored solution takes place in the measuring vessel into which the sensor part with mirror element is dipped (Fig. 3.6). The sensor part and the photometer part are connected to a glass fibre light guide. This type of construction offers the possibility of cleaning the mirror optics in a simple way, manually, mechanically or automatically with rinsing solutions.

The unfiltered light leaves the light source, passing through the first glass fibre bundle, then through the colored solution, and is reflected on the mirror element in such a way that it can be picked up by the second glass fibre bundle and passes via the filter element to the detector. The latter determines the incoming intensity of a pre-selected wavelength on the filter element. A curve filter is used as the filter element which can be continuously adjusted to 400–700 nm. According to the Bougner-Lambert-Beer law, in the absorption maximum a linear dependence between extinction and the concentration results (see Sect. 2.3.2). Measuring faults resulting from cloudiness and self-coloring of the sample or by ageing of the photometer lamp can be eliminated by means of a measuring program. Before the addition of the reagents the extinction A_1, and after the completion of color development extinction A_2 of the colored measuring solution are determined. Extinction A of the compound to be determined is then calculated as the difference and is taken for determining the sample concentration:

$$A = A_2 - A_1 = S \cdot c + B$$
$$c = \frac{(A - B)}{S}$$

where S is the increase of the calibrating plot, B is the blank value, and c is concentration of the parameter to be determined.

From this formula it can be seen that, for the determination of concentration, besides the analytical program in which the extinction of the sample is determined, the blank value B and the increase S of the calibrating plot must be determined at regular intervals. The measurement of blank value and standards can be carried out manually or automatically.

More recently, in process analytics of liquid systems, *IR spectroscopy* in the near infrared spectral range and *Raman spectroscopy* have been used. In the first

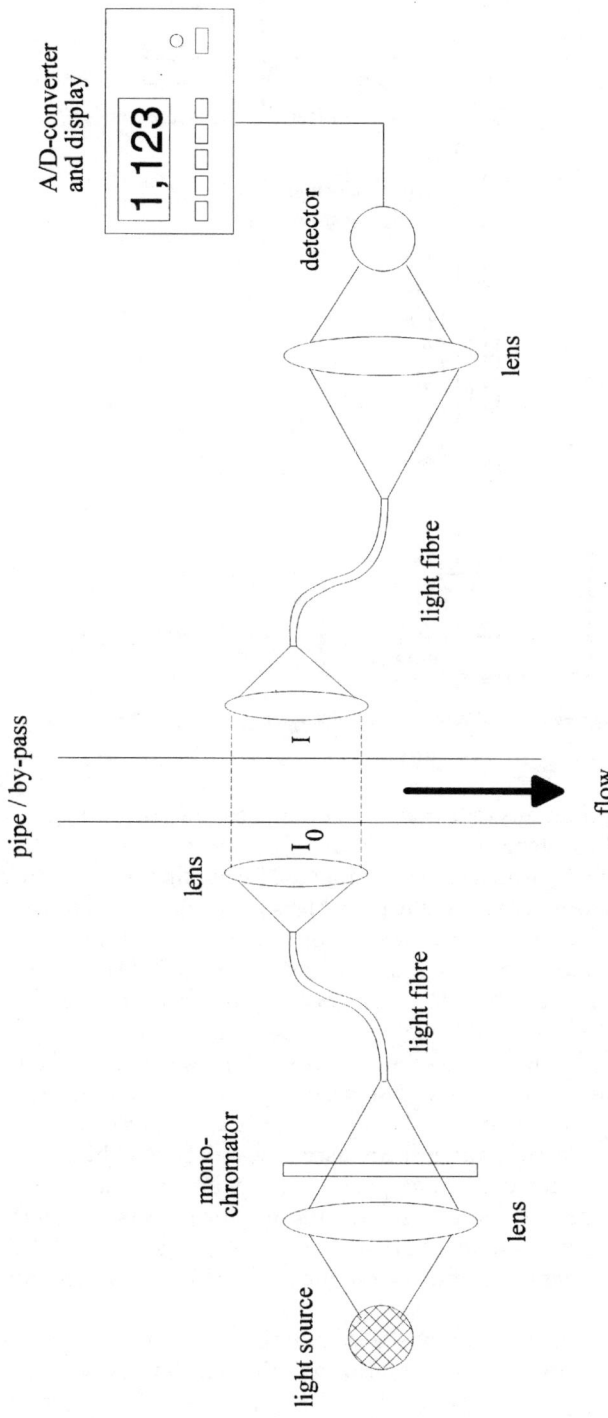

Fig. 3.5. Example of the application of light fibers in photometric process analytics

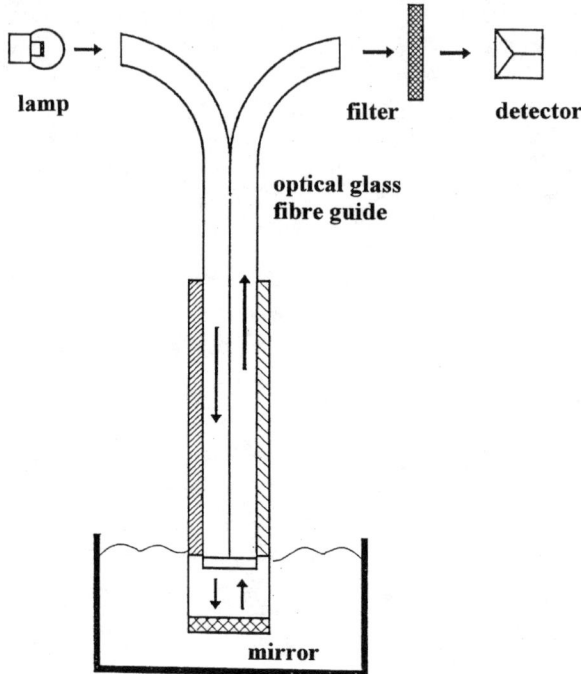

lamp

filter

detector

optical glass
fibre guide

mirror

Fig. 3.6. Sensor with mirror element of a process photometer

case, in which the components to be determined must show an absorption in the
near infrared range (wavelengths of 900–2400 nm), cuvettes in the process flow
can be installed externally which can be connected by light fibers of up to 200 m
in length to the spectrometer and make possible on-line measurement.

Raman spectroscopy, which in the past was only of very minor importance for
routine analysis, has more recently found wider application [21] by means of
new, low-price mini-spectrometer devices, new sampling techniques and the use
of light fibers [22]. Modern vibrational spectroscopic methods such as Fourier
transform infrared spectroscopy (FT-IR spectroscopy) as well as the FT-Raman
spectroscopy combine the advantages of a high information content, rapid provision of the information required and simple sample preparation. Until recently, however, the application of Raman spectroscopy was limited to basic research.
The sensitivity of traditional Raman spectroscopy with its visible laser light
sources against disturbances by fluorescence and photolysis was responsible for
this. Only since the development of NIR-FT-Raman spectroscopy (NIR=near
infrared) have the strong points of this versatile measuring method been recognized to the full.

Brief mention should now be made of the basis of Raman spectroscopy. If
molecules are excited with monochromatic radiation (e.g. lasers) the effects of
collision processes of the exciter light quanta with the molecules can be

observed. As a consequence of inelastic collisions, radiation arises from quanta which have transferred oscillation energy to the molecule or have absorbed oscillation energy from it. These oscillations can be characterized as stretching frequencies or deformation oscillations. In the case of stretching frequencies it is mainly the lengths of the links that change; in the case of deformation oscillations, on the other hand, it is the link angles. Characteristic bands are characterized by defined Raman (and infrared) intensities in certain frequency ranges. Since frequencies and intensities depend on all structural parameters, Raman spectroscopy can be used as a foolproof "finger print" for identifying a compound.

The radiation of the samples is effected in the case of FT-Raman spectroscopy with a near-infrared laser (1064 nm) which does not damage the sample compared to visible laser light. The interaction of the laser radiation with the sample

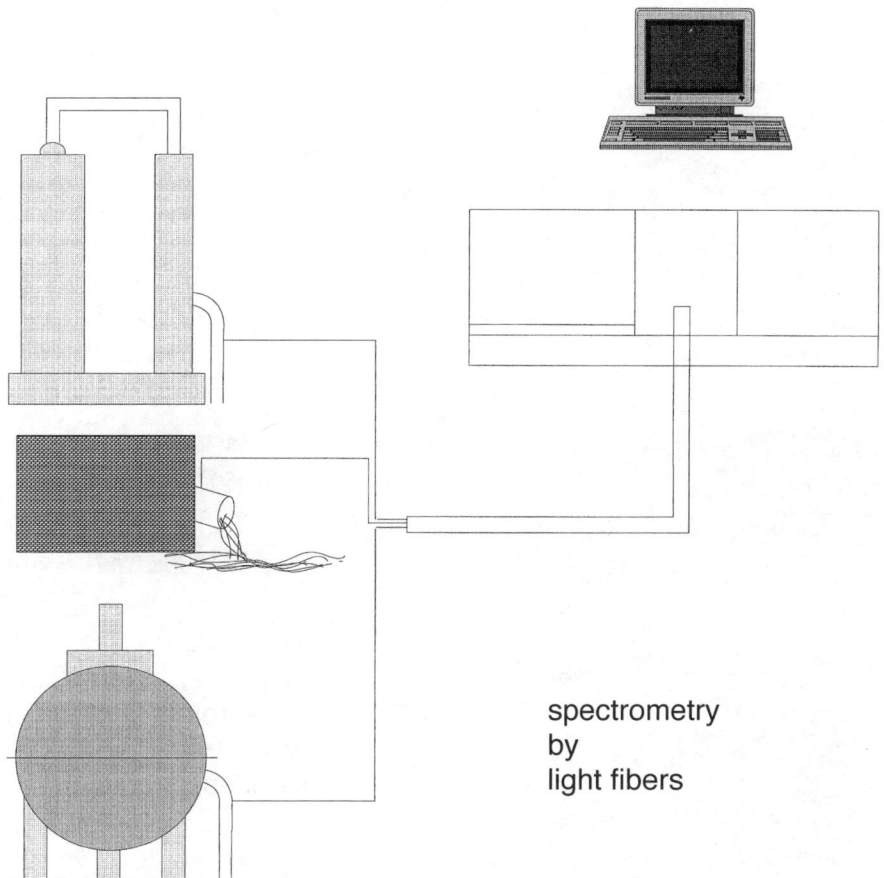

spectrometry
by
light fibers

Fig. 3.7 A: Scheme of spectrometric process control by light fibre technique

leads to the inelastic scattering process known as the Raman effect. The resulting Raman spectrum permits first of all statements about the structure of the
sample investigated. The clear allocation can be supported by a spectra search in
an FT Raman library. In the meantime, evaluation programs for establishing the
identity and cluster analysis as well as multivariate calibration methods are
available.

Seen from the application side, FT-Raman spectroscopy is not only an alternative to FT-IR spectroscopy but sometimes the only viable means of analysis.
The sample preparation as a rule is restricted to the simple insertion of the substance to be examined into the measuring device. Investigations can be carried
out, for example, directly in glass vessels or in the aqueous milieu. On the equipment side, this advantage is supported by a pre-adjusted sampling device or by a
light fibre sensor with which samples can be measured "on site". Thus interesting possibilities for production control and for environmental monitoring are
opened up [23] (Fig. 3.7).

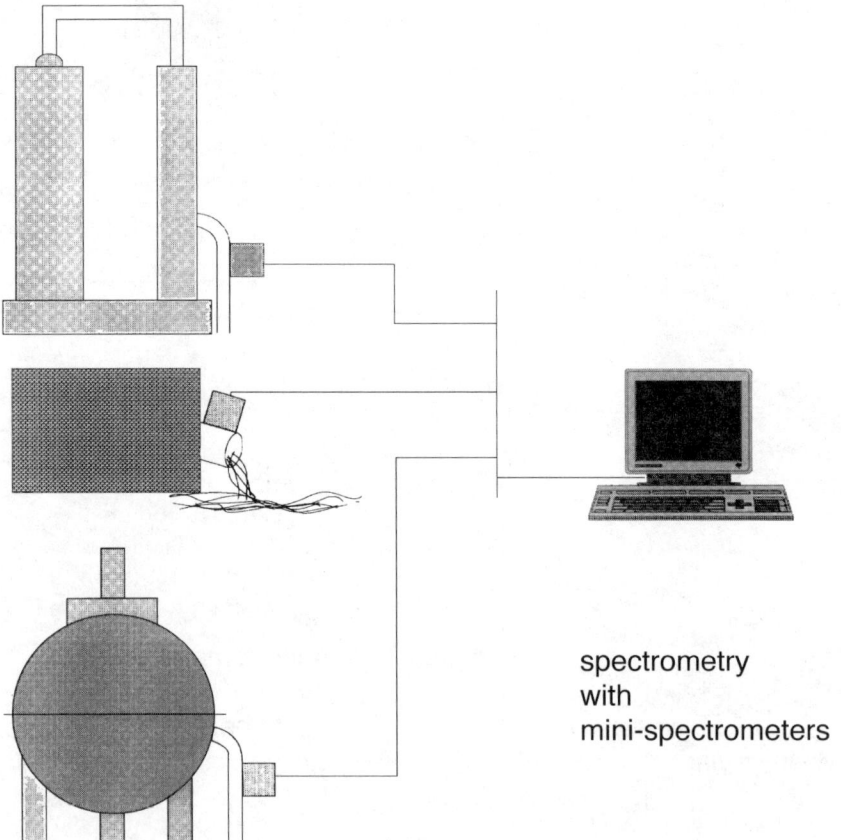

spectrometry
with
mini-spectrometers

Fig. 3.7 B: Scheme of spectrometric process control with decentralized mini-spectrometers

Table 3.4. Industrial examples for X-ray fluorescence spectrometric on-line analysis

Industrial section	Products	Analytes
Mineral oil refining	Crude oil, oil fractions, fuel oil, fuels	S, Pb, Mn
	Oil residues, crack products	V, Ni, Fe
	Lubricating oil production	Ca, Zn, P, S, Cl, Ba
	Waste water treatment	S, Cr, Pb, and other metals
Chemical industry and inorganic chemical engineering	Sulfonization and chlorination of hydrocarbons	S, Cl
	Extraction processes in hydrometallurgy	e.g. Cu, Cl, Hf, Zr
Galvanic industry	Pickling baths	Ag, Au, Cd, Cr
	Electrolytes	Co, Ni, Zn, P, S, ...

As a further spectroscopic analytical principle, X-ray fluorescence spectroscopy and X-ray diffractometry (see Sects. 4.2.1 and 4.2.2) have been introduced successfully into industrial on-line analysis. The significance and the width of application are shown by the examples listed in Table 3.4. As excitation sources, in these cases, various radioisotopes (see Sect. 4.4) are used. The supplier publications issued by various equipment manufacturers frequently represent suitable sources of information.

3.2.1.2
In-Line Analytics and Sensor Technology

As in-line analytics those methods are designated in which the determination of the analyte takes place directly in the production flow without sampling from the process medium. In that way, on the one hand sampling and sample preparation errors are avoided, on the other hand this way of proceeding makes possible a practically delay-free measurement and thus offers a possibility of almost directly influencing the process event. "Direct" analysis of a process flow necessitates probes or sensors which are specific to the component to be determined or provide a measuring signal characteristic of the current state of the process.

Methods for in-line analysis have been developed for matters of all three aggregate states. Selected examples for process analytics of melts and of solids are dealt with in later sections (see Sects. 4.3.5, 6.2.2, 6.3.2.5 and 6.5). Since, for many interesting process parameters, no sensors are yet available, process analytics will not be able to manage without on-line and off-line concepts. The sensors are mostly brought into contact with the process fluid in the form of a special probe ("*chemical sensor*" as miniaturized measuring device) which is able to convert information on a chemical substance (concentration) on-line into an

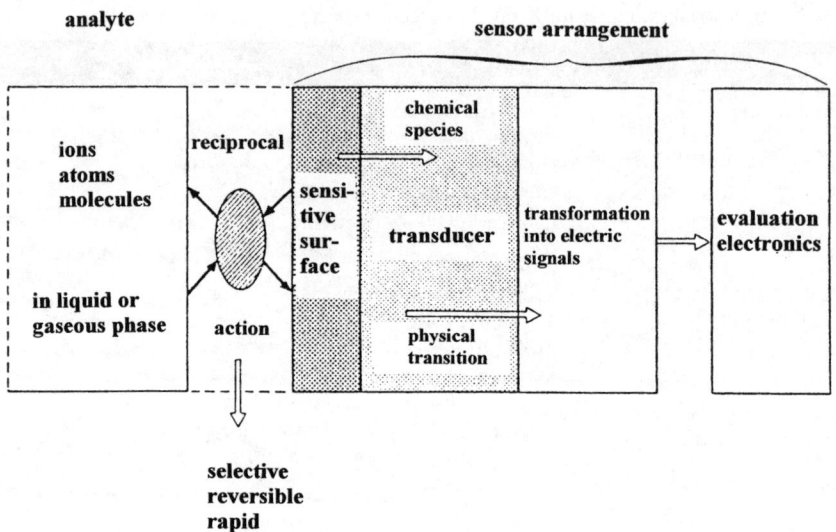

Fig. 3.8. Principle and reciprocal action of chemical sensors (schematically)

analytically useable signal [24, 25]. Thereby, both chemical reactions with the analyte and also its special physical properties within the overall system can be used [26].

Such a sensor in principle consists of three components. In the *receptor* part selective identification and, if necessary, the transformation of the substance takes place, while in the *transducer* part conversion into the actual measuring signal occurs (Fig. 3.8). Downstream of the transducer there is the measuring electronics in the form of an impedance converter or a measuring amplifier. Then the measured value output is obtained, e.g. by means of a data processing system.

Signal formation can be of a purely physical nature (e.g. change of the refraction index), or a specific molecular identification can take place by a defined chemical reaction (e.g. complexing in the case of ion-sensitive liquid membrane electrodes). Depending on the sense of sensor principle used [24] for this purpose, various transducers are available which are mainly based on the following methods [27]:

- electrochemical (voltametric, potentiometric sensors, chemically sensitized field effect transistors (chemical FETs));
- electric (metal oxide semiconductor sensors, organic semiconductor sensors, electrolytic conductivity sensors, dielectricity sensors);
- optical (measurement of absorption, reflection, luminescence, fluorescence, a refraction index, an optothermal effect, light scattering);
- mass-sensitive (piezoelectric devices, surface-acoustic-wave (SAW) devices) [28];

- magnetic (measurement of paramagnetism);
- thermometric (measurement of heat change).

Molecular identification based on biochemical processes (or reactions) leads to the *biosensors* defined as a sub-group of the chemical sensors (see Sect. 5.3).

A chemical sensor can ideally be an independent analytical unit. However, it can also, as an individual detection unit, be an essential component of an automated analytical system which enables additional work steps such as self-calibration or sample preparation. Thereby, the trend-setting use of sensorics is closely linked with the development in microelectronics and the application of chemometry for the optimization of sensor properties [29].

Electrochemical detection is carried out in the case of chemical sensors mainly potentiometrically, amperometrically or conductometrically. The potentiometric sensors [30] include, besides the pH glass membrane electrode which has been known for quite a long time, all other ion selective electrodes which can be sub-divided, depending on the type of construction, into glass membrane, solid state or liquid membrane electrodes. Combined with a constant-potential reference electrode to form an electro-chemical measuring chain the voltage of this measuring chain is dependent on the activity of the measuring ion in the analytical solution.

Such sensors working potentiometrically can, for example, be used in process analytics as well as for continuous monitoring of ground water and drinking water in the field of waste disposal sites or contaminated areas (e.g. the direct control of the waste disposal site for any leaks) [31]. Amperometric sensors are used mainly for the measurement of gases whereby conductometric sensors are particularly frequently applied in water examination as conductivity measuring cells for checking the electrolyte content. These sensors are, in fact, very robust and reliable, but they do not have any selectivity at all so that statements cannot be made about the type of electrolyte.

Optical measuring methods offer a large number of possibilities for the development of sensors which can be used in biological process analytics. Here the spectrum extends from on-line sensors working non-destructively and with rapid response on the basis of turbidimetric and spectral fluorometric methods up to (quasi) continuously working enzyme or antibody-aided intelligent biosensor systems [32]. The development of optical sensor systems is closely connected to progress in the field of glass fibre technology and integrated optics, the aspect of miniaturization playing an important part. Summing up, it can be stated that the possibilities for the use of already existing sensors have so far by no means been fully exhausted [33] and that sensors will gain increasing importance for process analytics.

3.2.1.3
Off-Line Concepts

Since process analytics does not only take place on-line or in-line locally on-site, but also off-line in central laboratories, there must also be taken into account the special requirements of the processes to be controlled. That, as a rule, means automation of the analytical sequences including data processing which, if necessary, includes automatic remote transmission of the analytical results.

These relationships between technical requirement and analytical problem solving can be shown by two examples from industrial practice.

3.2.1.3.1
Photometric Multi-Component Analysis

The first example deals with water and waste water investigation within the framework of water and circulation economy in an iron and steel works [34]. The requirement consists of the comprehensive determination of the anions (besides the cations) in this water on the basis of technical specifications. The efficient solution is achieved by the use of a photometric multichannel analyzer in which the well-known working steps of photometry are mechanized or automated.

In this case, it should be mentioned that the use of such an analyzer is only worthwhile from about 60 samples per day. The preliminary work (washing in, temperature stabilization, preparation of reagents) as well as the subsequent work (flushing of the overall equipment before shutting down) takes 30–60 min. In addition, double determinations are necessary if consecutive samples differ greatly in only one measuring variable. With the above-mentioned task the equipment must be constantly supervised since without re-adjustment of the zero points and without intermediate checking with standard solutions a series with more than ten samples should not be examined. With 60 or more samples per day for analysis with reference to four or more parameters there is, despite the operating work which should not be underestimated, a considerable time saving compared with the conventional way of proceeding. The analytical data thus determined can flow on-line into a data system which can be used for monitoring the overall water system.

3.2.1.3.2
Robotization of Analytical Work Sequences

In process analytical practice, however, it is necessary to analyze not only genuine solutions as in the above-mentioned case but also, in certain circumstances, emulsion systems which, already at the sampling stage, require chemical and analytical know-how. On this the following example should be described. It deals with an aqueous emulsion which contains 1–5% of organic components in fully demineralized water and is used as a rolling aid during the cold rolling of steel.

In this case, too, there is no possibility of on-line testing so that there is an urgent need to carry out a check as quickly and efficiently as possible on the production sequence of the day. Here, the solution consists of the use of the laboratory robot system [35] which shows a further possibility of "automation in analytical chemistry" [36].

The in-plant necessities will be dealt with later in more detail; here, the methodological approaches and the analytical solution of the problem are to be

Table 3.5.
Chemical composition of a
cold rolling oil

Component	% m/m
Mineral oil	30
Fatty acid and fatty acid esters	50
Emulsifier	5–10
Polar additives	1–2
S-additive	3–6
Antioxidants	<1
Corrosion inhibitors	<1

Table 3.6.
Concept of the analytical
control of a cold rolling
emulsion

Parameter	Statement
Oil content	Foreign oil / lubrification / cooling
pH	Cleanser introduction, bacterial attack
Conductivity	Foreign ions
Saponification value	Lubricating action / foreign oil
Fe	Dirt bearing capacity / abrasion
Si	Anti-sticking agent / influence on stability
Stability	Emulsifier efficiency
Cleaving property	Disposing

looked at. As can be seen from Table 3.5, rolling oils consist of a large number of functional ingredients, the presence and effectiveness in the emulsion of which are constantly controlled. Table 3.6 makes clear that, by determining the oil content, the pH value, the conductivity and the saponification value, the lubrication effectiveness of an emulsion is ensured as well as outside oil ingress, bacterial attack and penetration of outside ions being analytically identifiable. The scheme of the investigation sequences can be seen in Fig. 3.9.

The following detailed description of the way of functioning of a laboratory robot is intended to clarify the expenditure which is necessary to convert relatively simple analytical work steps into robotized form.

The use of a robot within the frame of process analytics leads to the following advantages:

- reduction of personnel expenditure;
- freeing of skilled personnel from monotonous routine work (personnel can then be used for work requiring higher qualifications);
- shifting to night time of the analytical work once carried out only during the day;
- protection of the staff from dangerous work substances;
- avoidance of person-related analytical errors;

Fig. 3.9.
Sequences of a robotized oil
emulsion investigation

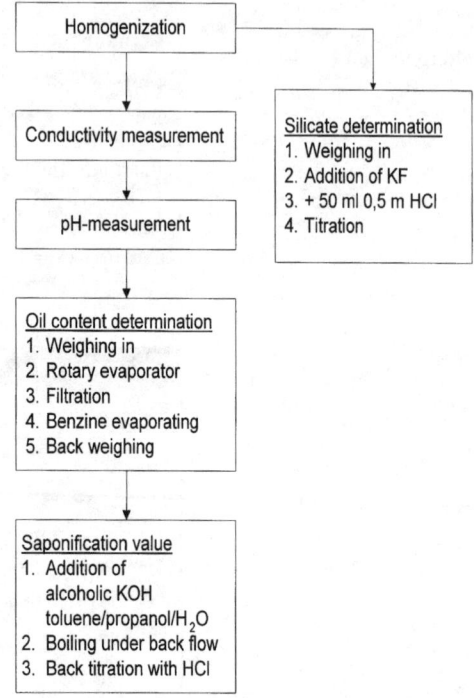

– coordinated linking of work cycles to samples of several emulsion systems;
– on-line output of data to the plants.

The conceptional structure of a robot system (Fig. 3.10) consists of the *robot* as such, a *transfer arm* with the possibility of moving to individual points in the work room very precisely, a *control unit* which

– starts up the arm,
– stores the coordinates,
– starts up/switches off or controls external work stations,
– determines pH value and conductivity via calibration functions,
– outputs analysis data,

and *work stations* which

– carry out mechanical or chemical operations.

With remote control the robot arm can be led into certain space co-ordinates which then, by designation with usual laboratory terms such as "pick up sample" or "lift piston", can be fixed in the program and exactly reproduced with the terms mentioned. By learning standard operations and combining various work sequences to a "standard term" by re-use of this "standard term" at any points in the sequence program, all work programs can be easily incorporated.

A specially shaped transport hand for large-volume sample bottles places, at the beginning of the analysis, the appropriate sample bottle for homogenization of the analysis solution onto a magnetic stirrer.

The suction cannula at the other end of the transport hand connected with tubing is, after pH value and conductivity have been directly determined in the sample bottle, used for transferring an aliquot sample quantity from the sample bottle into a round flask. The round flask is placed from the rack onto a top-mounted dish analytical balance, balanced and has about 50 g of sample solution poured into it. The absolutely measured weighing-in is taken over by the control unit and used for later calculation of the oil content.

For evaporation of the water portion in the emulsion the round flask is placed, by the robot arm, onto an automatic rotary evaporator. The flask is held by the robot until the vacuum is switched on, preventing the flask from slipping off. The whole evaporator upper part is then lowered until the flask is fully immersed in the water bath. The water level of the bath is fully automatically level-controlled. After the completion of the evaporation process the arm of the robot takes hold of the flask neck and carefully draws the flask from the Teflon-coated ground joint after the control unit has switched off the vacuum pump.

After the addition of normal benzine the oil/iron residue is removed on a horizontal shaker from the flask rim. Using filtration through filter wool into the ground joint flask provided for later saponification, the oil is separated from the solid content. The filtrate is freed of benzine again by blowing nitrogen through it and simultaneously heating the flask in an aluminum block. The remaining oil shows gravimetrically the oil content of the emulsion.

Fig. 3.10.
Schematic example of a
laboratory robot system

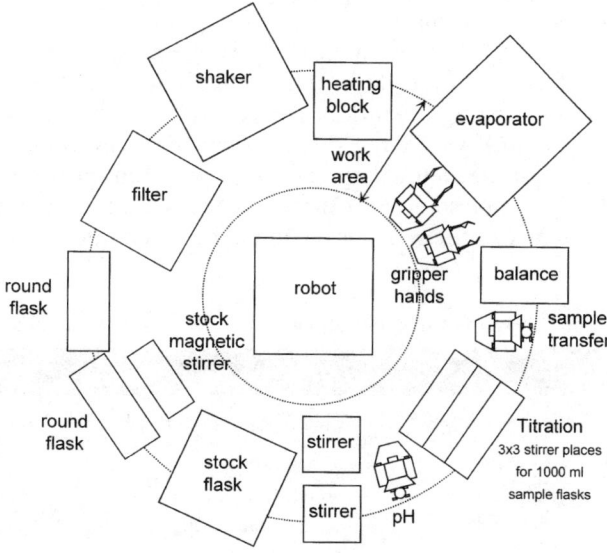

In order to carry out the saponification reaction with alcoholic potassium hydroxide solution it is necessary to fit a reflux condenser that is pneumatically movable on a holder above an aluminum block.

While the alcoholic potassium hydroxide solution is metered into the reaction flask on the balance because of the necessary precision, the required solvent of toluene, isopropanol and water is added through a nozzle at the top end of the reflux condenser. The saponification lasts 1 h.

Other work cycles such as retitration of the potassium hydroxide solution unused after the saponification reaction, titration for silicon determination and pH measurement are processes for which there are ready-made directly combinable modules available from manufacturers.

3.2.2
Melts

The melts of metals and oxide mixtures are, naturally, process analytically interesting and industrially important liquids. This subject will only be mentioned by examples while the overall problems posed in these cases will be dealt with in connection with the processes of pig iron and steel production (see Sect. 6.3.2.1.).

Besides the element iron (steel), the element aluminum is of considerable industrial importance so that a short look should be taken at the process analytical component connected with it.

Until just a few years ago, furnaces for electrolytic production of aluminum from alumina were still operated empirically [37]. For conception of controlling the process neither the necessary information from the furnace unit nor automatically working operation units were available. With the increasing possibilities of process control by data processing and automation, the demand for information on the operating conditions grew [38].

In the case of fusion electrolysis one must, on the one hand, deal with a melting flux electrolyte and, on the other hand, with the aluminum metal as a product for casting operation. For control of the melting flux electrolyte (chiolite $Na_5Al_3F_{14}/Al_2O_3$) X-ray diffraction and X-ray fluorescence analysis are used, while for the metal Al optical spark emission spectrometry is applied (Table 3.7). Ladle samples and special cylindrical samples are used, with the solidification structure of the samples having to be taken into account [39].

Table 3.7. Process analytics of the fusion electrolysis for aluminium production

Melt	Analyte	Analytics
Fusion electrolyte	NaF, AlF$_3$, CaF$_2$, LiF, Al$_2$O$_3$	Production of powder briquettes; XRD, XRF
Pure aluminium Metallic melts	Al (degree of purity) Al, accompanying elements	Pneumatic tube system; automatic sample preparation; OES

Table 3.8. Process analytics of the BAYER process for aluminium oxide production

Process matter	Analytical task	Direct measurement on site	Analytical method
Suspensions	Solid matter content (bauxite, red sludge, $Al(OH)_3$)	Radiometry	Gravimetry
Circulating lye	Na_2O and Al_2O_3 content	Conductivity	Automatic titration
Bauxite, red sludge $Al(OH)_3$, Al_2O_3	Moisture and loss at red heat	–	Gravimetry
	Elemental analysis	–	Powder briquettes and fusion decomposition; XRF

For the sake of completeness, it should be mentioned that the production of pure alumina [40], being used as the starting product and obtained by means of the BAYER process, is also subjected to continuous process analytics [41] (Table 3.8).

In all metal industries there are, at present, discussions [42] about the at-line installation of analytical devices near the melting units or automation of the analytical methods for the improvement of process control without any uniform recommendation so far having been made. This is above all due to the fact that the questions asked are very varied.

3.3
Suppliers for Process Analyzers and Off-Line Analytical Equipment for Process Control

3.3.1
Process Titrators

Bran & Lübbe GmbH, Werkstrasse4, D-22844 Norderstedt
Deutsche Methrom GmbH & Co, Postfach 1160, D-70772 Filderstadt
Horiba Europe GmbH, Industriestrasse8, D-61449 Steinbach/Taunus
Polymetron/Zellweger Uster GmbH, Bensberger Strasse 249, D-51503 Rösrath
Siemens AG, Meß- und Regeltechnik, Kruppstrasse2, D-45128 Essen
H. Wösthoff GmbH, Max-Grewe-Strasse 30, D-44727 Bochum
Wissenschaftlich-Technische Werkstätten GmbH (WTW), Postfach 1642, D-82360 Weilheim

3.3.2
Electrochemical Analyzers

Deutsche Metrohm GmbH & Co (see above)
Knick Elektronische Geräte GmbH & Co, Beuckestrasse22, D-14163 Berlin

Orbisphere GmbH, Postfach, D-35337 Gießen
Pfaudler Werke GmbH, Postfach 1780, D-68707 Schwetzingen
Polymetron/Zellweger Uster GmbH (see above)
Radiometer GmbH, Am Nordkanal 8,D-47826 Willich
Siemens AG (see above)
Dr. Thiedig & Co, Anlagen- und Analysentechnik, Lausitzer Strasse 10, D-13357 Berlin
WTW (see above)
Yokogawa Electrofact, Berliner Strasse 113, D-40880 Ratingen

3.3.3
Process Photometers

Alliance Instruments, Hugenottenstrasse111, D-61381 Friedrichsdorf
Bran & Lübbe GmbH (see above)
Deutsche Methrom GmbH & Co (see above)
Horiba Europe GmbH (see above)
Maihak AG, Postfach 601709, D-22292 Hamburg
Perkin-Elmer Bodenseewerk GmbH, Postfach 101761, D-88647 Überlingen
Polymetron/Zellweger Uster GmbH (see above)
Polytec GmbH, Polytec-Platz 5–7, D-76337 Waldbronn
Siemens AG (see above)
Tecator/Perstorp Analytical GmbH, Postfach 1370, D-63085 Rodgau
WTW (see above)
Carl Zeiss Jena GmbH, Tatzendpromenade 1 a, D-07745 Jena

3.3.4
X-Ray Fluorescence Spectrometer/X-Ray Diffraction Spectrometer

ASOMA Instruments Vertriebs GmbH, Ingoldstädter Strasse 68 a, D-80939 München
EG & G berthold / Laboratorium Prof. Dr. Berthold GmbH, Calmbacher Strasse 22, D-75323 Bad Wildbad
Outokumpu KM-Analytik GmbH, Königsteiner Str.98, D-65812 Bad Soden
Österreichisches Forschungszentrum Seibersdorf GmbH, A – 2444 Seibersdorf
Philips Industrial Electronics Deutschland GmbH, Miramstrasse87, D-34123 Kassel
Siemens AG, Röntgenanalytik, Postfach, D-76181 Karlsruhe
UNICAM Analytische Systeme GmbH, Korbacher Strasse 75–77, D-34132 Kassel

3.3.5
FT-IR/Raman Spectrometer

Laboratorium Prof. Dr. Berthold GmbH & Co KG, Calmbacher Strasse 22, D-75323 Bad Wildbad
Bio-Rad Laboratories GmbH, Bischofstrasse86, D-47809 Krefeld

Bruker Analytische Meßtechnik GmbH, Silberstreifen, D-76287 Rheinstetten
Nicolet Instrument GmbH, Senefelder Strasse 162, D-63069 Offenbach
Perkin-Elmer Bodenseewerk GmbH (see above)

3.3.6
Laboratory Robots

ISRA Systemtechnik GmbH, Industriestrasse14, D-64297 Darmstadt
Zymark GmbH, Black & Decker Strasse 17, D-65510 Idstein

3.3.7
Fluorescence/Stray Light Sensors

Ingold Meßtechnik GmbH, Siemensstr.9, D-61449 Steinbach
Monitek GmbH, Moersenbroicher Weg 200, D-40470 Düsseldorf

3.3.8
Remarks

The preceding compilation cannot make any claim to completeness. For further
information reference should be made to the literature and the market surveys
published in "Nachrichten für Chemie, Technik und Laboratorium", the news of
the Gesellschaft Deutscher Chemiker (GDCh) (Society of the German Chemists).
Further information is given in Sects. 4.5 and 6.9.

References

1. Kaiser R (1966) Fresenius' J Anal Chem 222:128
2. Melzer W Jaenicke D (1980) In: Ullmanns Encyklopädie der technischen Chemie, 4. Aufl.,
    Bd. 5, Verlag Chemie, Weinheim, p 891
3. Brüggemann B, Kudermann G, Lührs C (1987) Erzmetall 40:120
4. Surmann P (1988) Galvanotechn 79:3287
5. Kairies P (1990) Techn Mitt 83:H.4 :205
6. Jones JG (1985) ESN-Eur Spectr News 63:34
7. Nölte J (1990) Galvanotechn 81:2738
8. Bolch T (1987) Galvanotechn 77:3558
9. Gantner E, Steinert D (1990) Fresenius' J Anal Chem 338:2
10. Hönig U (1989) Metalloberfläche 43:398
11. Brandstetter H, Busch E, Maser N, Müller F, Wiedemann T (1987) QZ Qualität und Zuver-
    lässigkeit 32:8
12. Frenzel F, Selan M (1992) GIT Fachz Lab H 12:1239
13. Möller J (1988) Analytiker-Taschenbuch, Bd. 7, Springer, Berlin Heidelberg New York, p 199
14. Müller F (1985) CAV 1
15. Müller F (1992) Kontrolle Dezember, 6
16. Burkhardt W (1988) Galvanotechn 79:3252
17. Dahms W (1989) Metalloberfläche 43:345
18. Emmrich R, Fiedler F (1994) Labor Praxis Mai, 42
19. Danigel H, Gross H, Zumbrunn W (1989) Techn Messen 56:285
20. Schlemmer H, Katzer J (1987) Fresenius Z Anal Chem 329:435
21. Schrader B (1989) GIT Fachz Lab H 10:981

22. Schrader B (ed) (1994) Infrared and Raman Spectroscopy – Methods and Applications. VCH Chemie, Weinheim
23. Schrader B (1992) GIT Fachz Lab H 2:96
24. Nießner R (1988) In: Analytiker-Taschenbuch, Bd. 7, Springer, Berlin Heidelberg New York, p 55
25. Cammann K, Ross B, Hasse W, Dumschat C, Katerkamp A, Reinbold J, Steinhage G, Gründig B, Renneberg R (1994) In: Ullmann's Encyclopedia of Industrial Chemistry, vol B 6, VCH, Weinheim, p 121
26. Göpel W, Hesse J, Zemel JN (eds) (1992) Sensors: a comprehensive book series in 8 volumes. VCH Chemie, Weinheim
27. Göpel W (1993) Nachr Chem Tech Lab 41:332
28. Wohltjen H, Dessy R (1979) Anal Chem 51:1458
29. Drobe H, Drost S (1992) GIT Fachz Lab H 10:1 034
30. Cammann K (1990) In: Neumer H, Heller W (eds) Untersuchungsmethoden in der Chemie. Thieme, Stuttgart
31. Hasse W, Ginter B, Cammann K (1992) GIT Fachz Lab H 6:6631
32. Anders KD, Plötz E, Scheper T (1991) GIT Fachz Lab H 10:1093
33. Cammann K, Lemke U, Rohen A, Sander J, Wilken H, Winter B (1991) Angew Chem 103:519
34. Sebastiani E, Loose W, Koch KH (1979) Stahl und Eisen 99:1487
35. Sommer D, Koch KH, Grunenberg D (1988) GIT Fachz Lab 32:1075
36. Malissa H (1986) Mikrochim Acta II:3
37. Wilkening S (1986) PdN-Ch 35:21
38. Mannweiler U, Schmidt-Hatting W, Koutny M (1983) Erzmetall 36:274
39. Kudermann G (1989) In: Schriftenreihe der GDMB Gesellschaft Deutscher Metallhütten- und Bergleute, Heft 55
40. Lotze J, Winkhaus G (1986) PdN-Ch. 35:18
41. Gado P, Szalay I, Farkas F, Feher I (1983) In: Light Metals 1983, Proc 112th AIME Annual Meeting, Atlanta
42. Analytische Schnellverfahren im Betrieb (1989) Schriftenreihe der GDMB Gesellschaft Deutscher Metallhütten- und Bergleute, Heft 55

4 Process Analytics of Solid Materials

4.1
Sampling of Product Flows

In the chemical industry, process analytics for solids plays a smaller part compared with that for gases and liquids. By comparison, this sub-area of analytics has, in some other sectors of industry, a not inconsiderable importance, as will be shown by some examples. Since solid product flows, as a rule, show spatial inhomogeneity, this point has to be taken into account during sampling. This can, for example, be achieved by the mixing of sample flows taken off at various points or by intermittent taking of samples from the total product flow at certain intervals. Thus, the number and the mass of increments is of essential importance for the assessment of the result. There are a number of standard approaches to the dependence of the mass of an increment on the particle size (Table 4.1) which directly reveal these difficulties [1].

A mechanical (automatic) sampling system for solids which can work either time- or weight-proportionally [2] must meet certain conditions in order to be able to exclude systematic errors.

1. The product flow must be traveled through by the sampling device at regular intervals.
2. The sampling device must evenly cover the whole cross-section of the product flow at a constant speed.
3. The opening of the sampling vessel must have a width which is at least three times that of the diameter of the largest grain in the product to be sampled in order to give each part of the material the same chance to become part of the sample.

Table 4.1.
Dependence of increment mass from particle size; example: iron ore (ISO 3081)

Grain size (mm)	Minimum mass of increment (kg)
150– 250	40
100– 150	20
50– 100	12
20– 50	4
10– 20	0.8
below 10	0.3

4. The sampling vessel is to be dimensioned so large that it can take up the increment taken in during travel through the product flow and that it does not overflow even if the flow rate is briefly exceeded.
5. The speed of the mechanical sampler with which it moves through the product is to be adjusted in such a way that as little spray grain, i.e. product deflected by the sample vessel, as possible is produced.

Similar conditions also apply to the conditioning following the sampling step, in which, after crushing, dividing of the product takes place. A representative part will, at the same time, be further processed while the remaining material is fed back into the product flow as rejected material. Particular attention is to be paid to the dividing stages. It has been shown that errors can easily occur during dividing of the samples. The grain size distribution of the reject portion must correspond to that of the separated sample.

Primary sampling from a product flow can be carried out in different ways. There is sampling:

– from falling product flow;
– from intermittently discharged product;
– from the moving belt;
– from the stagnant belt.

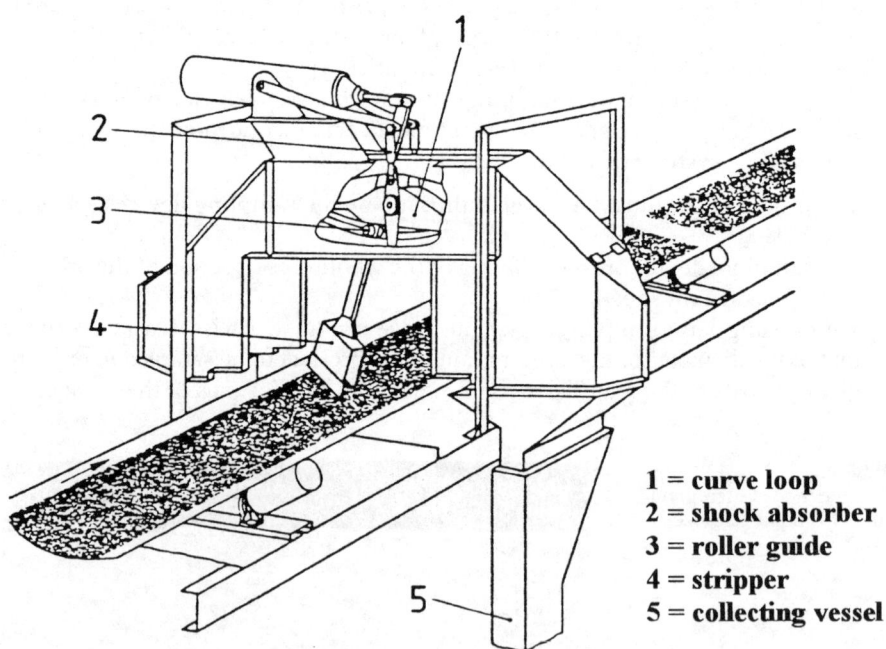

1 = curve loop
2 = shock absorber
3 = roller guide
4 = stripper
5 = collecting vessel

Fig. 4.1. Automatic sampling from a moving conveyor belt according to the German standard DIN 51701, p. 4 (by permission of DIN Deutsches Institut für Normung e.V., Berlin)

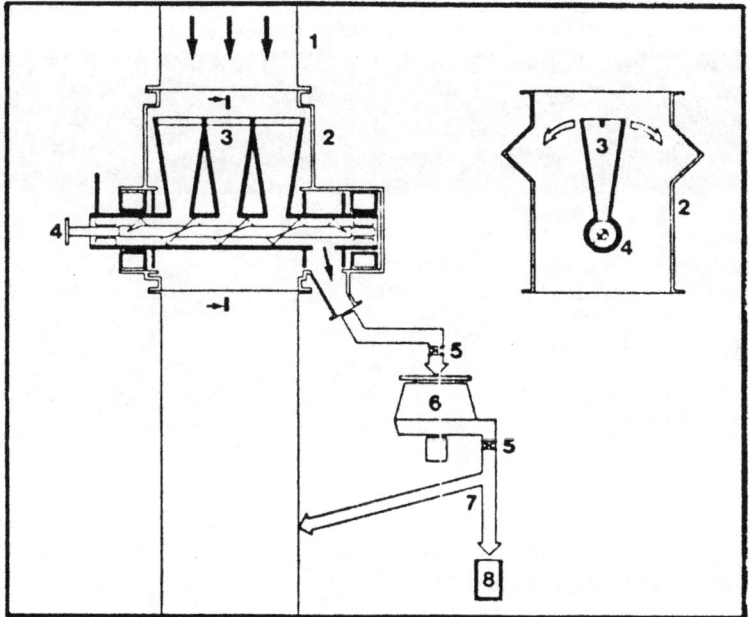

1 materialflow
2 built-in chute with bulk material inlet and outlet
3 pivoted sampling system
4 screw conveyer with increment transport
5 closing device resistant to pressure
6 sample mixer
7 sample divider with feedback system
8 laboratory sample

Fig. 4.2. Automatic sampling system for powdery materials according to the German standard DIN 51701, p. 4 (by permission of DIN Deutsches Institut für Normung e.V., Berlin)

The types of sampling equipment for these various conditions vary quite a lot and, for example, have been described in standards for the testing of solid fuels [2]. Figures 4.1 and 4.2 show two approaches to solutions for the sampling of solid mass products, the sampling of increments having to take place not only as a function of the mass but also of the homogeneity of the product to be sampled (see Table 4.2) [3].

For the sampling of solid matters in process analytics, there are, on the one hand, product flows of raw materials or mixtures of various substances to be used which flow to a process and, on the other hand, homogenized substance flow coming out of a process (intermediate or final product). The solids to be analyzed can be present both as "dry" (powdery, granular, lumpy) mass and as suspensions. For both forms examples are dealt with below. First of all, some analytical methods are to be briefly described which have proved successful in solids analytics for many years.

Table 4.2. Dependence of the number of minimum increments from quantity delivered and homogeneity of the material to be sampled: example – iron ore

Quantity delivered 1000 t	Number of increments dependent from the scattering of iron content		
	Large and the degree of homogeneity [a] >2% rel.	Medium 1.5–2% rel.	Small <1.5% rel.
70–100	180	90	45
45– 70	160	80	40
30– 45	140	70	35
15– 30	120	60	30
5– 15	100	50	25
2– 5	70	35	20
1– 2	50	25	15
0.5– 1	40	20	10
below 0.5	30	15	8

[a] Standard deviation of the iron content

4.2
Analytical Methods for Solids

4.2.1
X-Ray Fluorescence Analysis

The oldest example of process analytics of oxidic solid matters is represented by the automatic monitoring of raw meal composition during cement production [4]. In this case, the chemical analysis is performed, as in a number of other processes, by X-ray fluorescence spectrometry [5]. The results of the analysis are taken over by a process computer and are used as control variables.

The possibilities of X-ray fluorescence analysis (XFA) were, compared with other spectroscopic analytical methods, recognized and utilized at quite a late stage. Only when electronic measuring technology and data processing had developed far enough did the first useable analysis devices come onto the market in the early 1950s. The methods used for chemical analysis are based on the following principle: by radiation with gamma rays or X-rays the atoms in the sample to be investigated are excited to produce a characteristic, secondary X-ray radiation (fluorescence radiation), which at a certain angle (Bragg's equation) is reflected, its intensity and spectral distribution being examined by a detector system (Fig. 4.3). Since radiation of characteristic wavelengths, which can be separated by an analyzer crystal of known interplanar spacing, is attributable to each element, this method is suitable for chemical analysis [6].

The intensity of the fluorescence radiation of each element emitted is a measure of its amount in the sample. A distinction is made between the wavelength-dispersive XFA outlined here and the energy-dispersive XFA in which the X-ray fluorescence radiation is evidenced by a loss of energy in a semiconductor detector proportional to its energy.

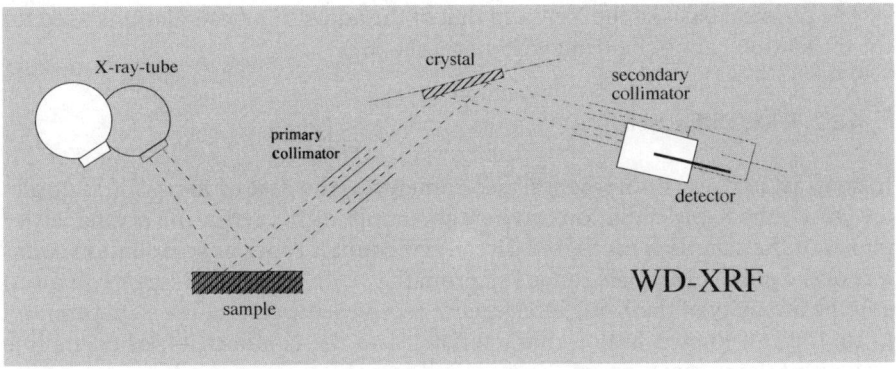

Fig. 4.3. Principle of wavelength-dispersive XRF

The quantitative evaluation of X-ray spectra is a complicated approach, since – apart from the analysis of thin films of about 10 μm – there is no proportional relationship between X-ray intensities and the composition of the sample. Here, interelement effects play an important role. In order to detect the influence of these effects and to be able to describe the relation between X-ray intensity and sample composition, experimentally and theoretically justified methods have been developed [6].

For the purposes of process analytics, in the case of wavelength-dispersive XFA simultaneous or multi-channel spectrometers are preferably used as they are able to measure e.g. 30 or even more different element lines simultaneously.

The precision and accuracy of XFA depends decisively on the nature of the sample. Thus, for example, in the case of compact or pelleted samples the radiated surface must be extremely plane and homogeneous. The grain size of the powdery material used in the case of oxides decisively influences the homogeneity of the sample surface and thus the radiation intensity. Moreover, the grain size distribution and the crystal structure of the sample matrix influence the measuring result. Therefore frequently fusion with a flux (e.g. $Na_2B_4O_7$, $Li_2B_4O_7$, $K_2S_2O_7$) is carried out in order to eliminate the influence of the original crystal structure and to reduce disturbing interelement effects by dilution. For calibration either (certified) reference materials, which have precisely known chemical compositions, or synthetic mixtures which correspond to the matrix of the material to be analyzed [7], are used.

With process analytical use of XFA, e.g. for the analysis of suspensions which in their original state flow continuously through a measuring cell, there is no defined sample. Correction of the grain size distribution, a homogenization of the sample or standardization of the matrix are not possible. Material must be investigated in the form in which it flows into or out of the process. Therefore calibration of these methods is far more problematic and the precision is markedly poorer than in the case of a defined, prepared sample. In pulp streams, because of the particular "sample geometry", only the elements above the ordinal number 17 are determinable with adequate precision.

The physical basis of the XFA and that of the adjustment calculations used for the evaluation cannot be dealt with here (see [6]).

4.2.2 X-Ray Diffraction

In contrast to X-ray fluorescence spectrometry, in the case of X-ray diffractometry (XRD) the X-ray radiation leaving the sample, diffracted at the crystal lattice planes of the sample is measured directly by counter tubes or scintillation counters over a particular angle range [8]. From the angle position (Bragg's equation) and the intensity of the X-ray interference received, the crystalline structure, and in the case of known lattice characteristic data the composition of crystalline solids and the percentages of their crystallographic phases, can be quantitatively determined [9]. A set of refraction angles and relative intensities is characteristic of each individual chemical compound. By comparison of data measured on samples of unknown composition with the values found in card files or data banks (e.g. ASTM card file, ICDD data bank), identification of the components can be carried out. Such data banks contain, in certain circumstances, line sets of more than 60,000 different crystallographic phases. For calibration, mixtures of known chemical and crystallographic composition are examined.

The determination of the different modifications of metals and alloys or the analysis of the various clinker phases during cement production are examples of the industrial use of XRD. As the width of the refraction lines is characteristic of the size of the crystallites in the sample, XRD can also be used for the investigation of the crystallinity and thus for the control of production steps in which crystallinity is a decisive variable.

A further field of application for X-ray diffractometry is the determination of the crystal structure of various phases. These examinations are, however, not restricted only to compact sample pieces or powder samples. Rather it is the case that analyses of thin films (5–500 nm) can also be performed. Besides phase analysis and the determination of the crystallite size it is possible to determine quite precisely and non-destructively the thickness, physical density and surface roughness of thin layers. These examinations of thin layers are, for example, of particular interest in the glass, metal and semi-conductor industries.

In the semi-conductor industry, for example, the lattice defects being linked with it, the stoichiometry of epitaxial layers is an important subject of investigation [10].

4.2.3 Neutron Scattering

In process analytics, radionuclides are successfully used, the radiation of which can be technically utilized in a large number of ways. These nuclear methods have the advantage not only of contact-free measurement but also of independence from temperature and state of aggregation. In particular for the determination of water in the most varied bulk goods, neutron scattering has been established as a measuring principle. An americium/beryllium radiation source

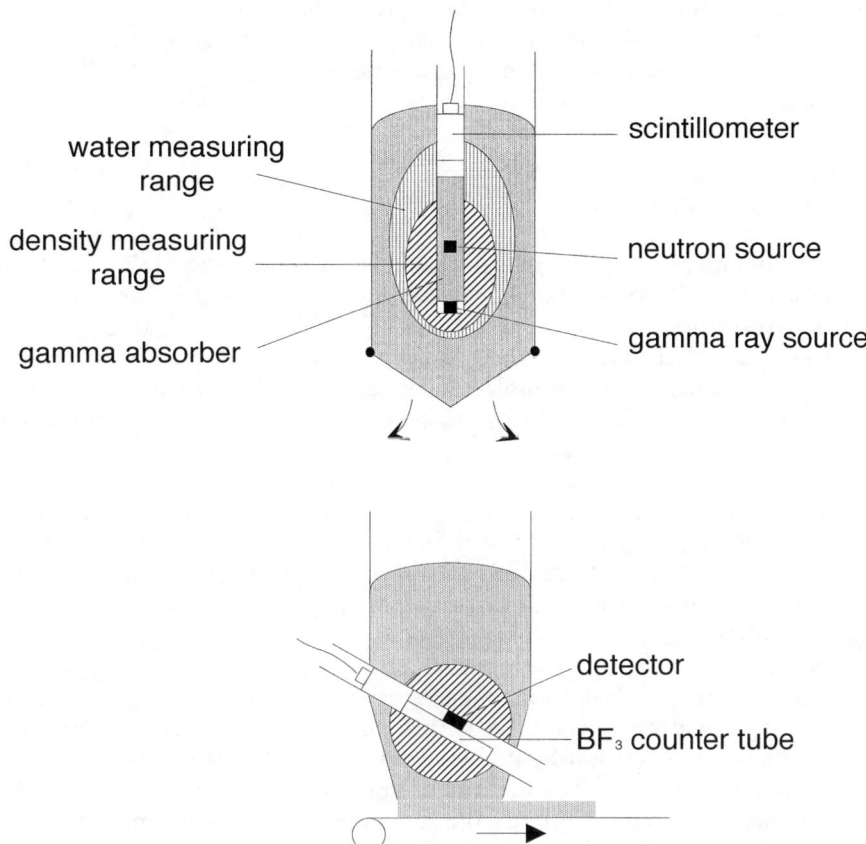

Fig. 4.4 A,B. Probes for the determination of water content of solid matters by neutron scattering

(100–300 mCi) emits rapid neutrons which are slowed down by scattering at the equally heavy hydrogen nuclei but which in the case of scattering at atom nuclei of higher mass hardly lose any speed [11, 12]. A shielded detector separated spatially from the source measures the number of neutrons after their interaction with the measured material (Fig. 4.4). When a fast neutron impinges on a hydrogen nucleus it transfers upon this impact more than half of its kinetic energy to the hydrogen nucleus, i.e. it is slowed down. After several impacts of this kind its kinetic energy has been reduced to thermal energy. Therefore around the source of the fast neutrons, in the presence of hydrogen (water) within the surrounding space, a cloud of slow neutrons forms, the density of which depends on the concentration of the hydrogen nuclei in the surrounding medium.

A detector for slow neutrons shielded from the source measures the number of particles which are a measure of the hydrogen or water content. The pulses measured per time unit are volume-wise proportional to the absolute water content. In order to obtain the water percentage as a percentage by weight, it has to be divided by the associated density of the material, e.g. coke, sintered raw mixtures, sand.

4.2.4
Microwave Transmission

As a further measuring principle for the moisture content determination of bulk materials, the following arrangement has proved successful in the most varied industrial sectors:

The product to be measured, e.g. on conveyor belts, is radiated with microwaves [12]. Thus, the free water molecules are set in rotation which causes a phase shifting and attenuation of the irradiated microwaves. Both effects are used for water content determination.

Instead of a fixed frequency, a wide frequency band can be used. That has the advantage that in each measuring cycle the phase shifting and attenuation are measured at a large number of individual frequencies and subjected to a processor-controlled plausibility analysis. In this way, disturbing resonance and reflection influences which can occur in the case of difficult measuring geometries can be suppressed.

Depending on the application, evaluation of the phase shifting and/or of the attenuation are used. In most cases of application high accuracy in moisture determination is achieved with a simple phase measurement, as the phase measurement is hardly influenced by material parameters such as temperature and grain size. The method of microwave transmission has, in comparison with reflection methods, the advantage that the whole material cross-section in the area penetrated by radiation is evaluated. Also in the case of inhomogeneous distribution of the water a representative measurement is thus possible.

4.3 Examples for Industrial Applications

4.3.1 Process Control of Cement Production

The dressing of the raw meal, regarding the achievement of an adequate homogeneity and a certain chemical composition, has a decisive influence on the cement quality. Therefore, in the cement industry, for the achievement of a homogeneous quality and optimum operating conditions, automatic processes for mixing control are used in which the analytics are an integral part of the control circuit (Fig. 4.5) [4].

From the raw meal flow a sample flow of about 60 kg/h is removed by a screw sampler from which, after tapering, laboratory samples are obtained at certain time intervals. For the elimination of grain size effects, fine grinding of the sample

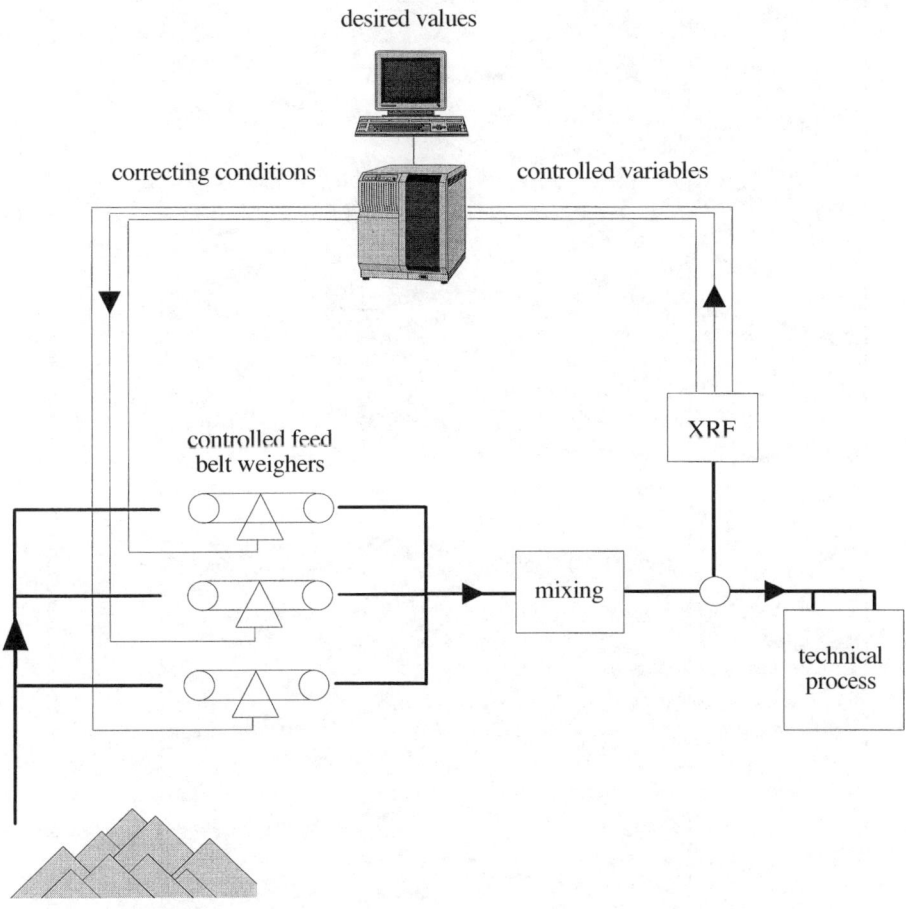

desired values

correcting conditions   controlled variables

controlled feed
belt weighers

XRF

mixing

technical
process

stock of raw materials

Fig. 4.5. Closed loop for the optimization of raw meal processing in the case of cement production

material with the addition of grinding aids is necessary (see Sect. 4.2.1). After that a pellet is made and analyzed by X-ray fluorescence spectrometry for the elements important for cement production Si, Al, Fe, Ca, Mg, K, Na, S, Cl. The analysis data for Ca, Si, Al and Fe flow into the mathematical model for raw meal control. From a comparison of desired values with the actual shares determined, adjustment variables are then obtained which become correcting conditions for the feed belt weighing integrated into the mass flows of the raw materials. By the use of XFA for raw meal homogenization (Fig. 4.6) the quality of the final product is decisively improved and the meeting of the quality criteria is guaranteed.

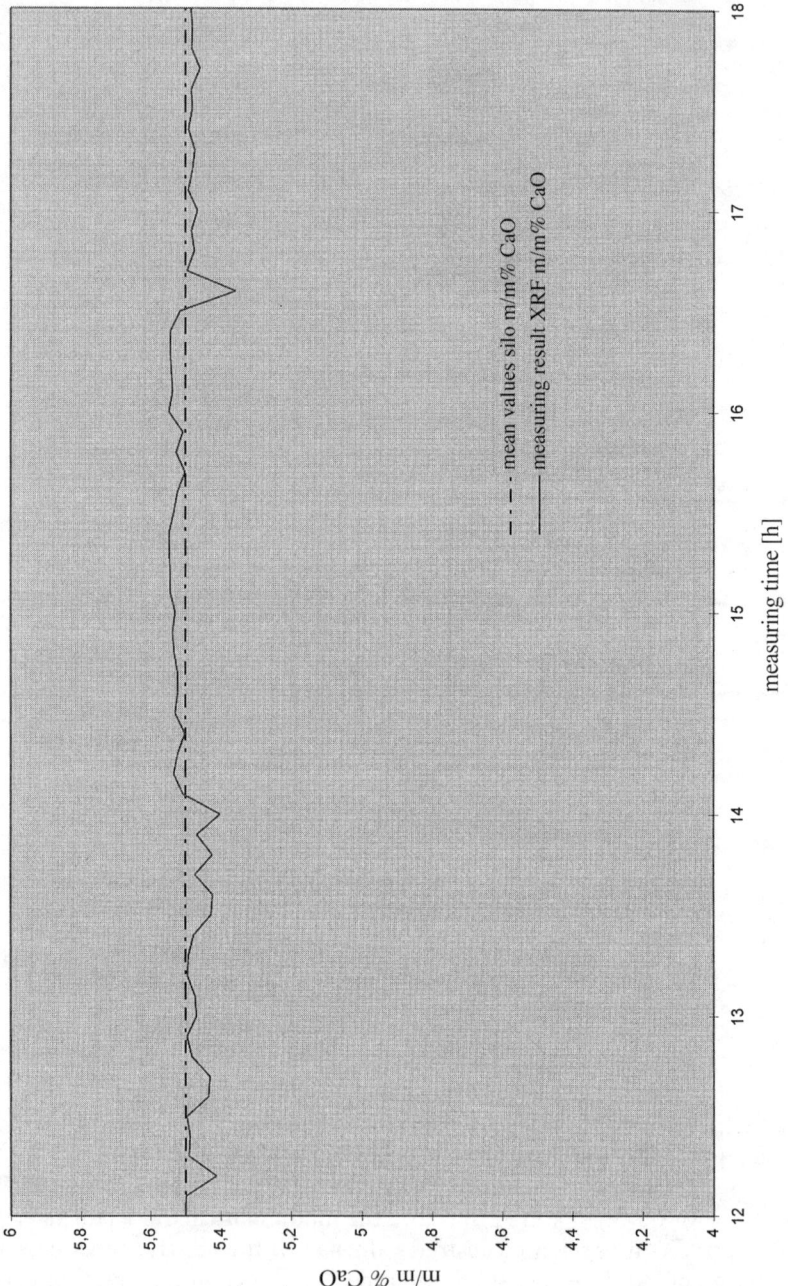

Fig. 4.6. CaO-content of cement raw meal after homogenization as a consequence of on-line analytics

4.3.2
On-Line Analysis of Flotation Processes

The overwhelming majority of the on-line X-ray fluorescence analyzers installed in industry are used for controlling flotation processes. For optimized monitoring of flotation units it is necessary to know the valuable substance content at the important operating points continuously and as precisely as possible. In this way, various optimization goals such as high yields and highly enriched concentrates can be decisively supported. While in the case of cement production (Sect. 4.3.1) powdery product has to be analyzed, the solids to be investigated in flotation processes are available in the form of suspensions which continuously flow to the analyzer as a branch current.

The solution of a process-analytical problem can consist of installing for each "process flow", slurry from one process stage, a separate measuring point. Another possibility is the alternating flow through a measuring device of samples of various origins with the help of a programmed control unit. For the flotation of *lead/zinc ores* the first approach mentioned has been chosen [13]. Here, the particular sample flow is measured quasi continuously with a 20 s measuring cycle. From the process flow a part current, e.g.120 l/min, is removed and passed through the sensing element of the X-ray fluorescence spectrometer. For the spectroscopic excitement a ^{241}Am radiation source is used. The various operating conditions of the analyzer including flushing, emptying and recalibration are adjusted and monitored computer-controlled.

During the flotation of lead/zinc ores the concentrations of Pb, Zn and Fe in the raw material and colliery tailings are of importance. Calibration of the measuring unit takes place with chemically analyzed samples of the various process stages.

By changes in the grain size distribution and the mineralogical composition of the raw material the precision and trueness of the measurements are considerably influenced. Therefore the calibration naturally applies only to the same conditions with regard to ore type, degree of grinding etc. The contents to be determined and the achievable reproducibility (precision) form Table 4.3. Figure 4.7 shows the X-ray spectra of samples with different zinc percentages [13].

The possibilities of the energy-dispersive XFA with radionuclides as radiation sources (Table 4.4) and X-ray diffractometry can also be used in further process analytical tasks to be carried out on solids for the optimization of flotation processes. Thus, the economical dressing of *sylvinite* (KCl, containing NaCl) as one of the most important potassium minerals requires the continuous determination of the KCl and NaCl contents of the solids in the pre-processing slurries. This problem can be solved by X-ray diffractometry [13] (Fig. 4.8). Further examples of the application of XRD are process control for the dressing and enrichment of *apatite ores* ($Ca_5[F, Cl, OH, 1/2 CO_3/(PO_4)_3]$) and the flotation of *talcum* ($Mg_3[OH_2/Si_4O_{10}]$) [14]. *Feldspar* and *quartz* as initial products for the glass, paper, paint and ceramic industries are also enriched by flotation. Pegmatite obtained by mining (magmatic rock containing feldspar, quartz and glimmer) is crushed and ground and then feldspar and quartz are separated by flo-

Table 4.3. Repeatability of the zinc, lead, and iron determination in lead/zinc slurries

Element	Raw material feeding (%)	Raw material feeding repeatability (s) [%]	Tailings (%)	Tailings repeatability (s) [%]
Pb	0–5	±0.10	0–0.5	±0.02
Zn	0–10	±0.25	0–2.0	±0.05
Fe	0–2	–	0–2.0	±0.05

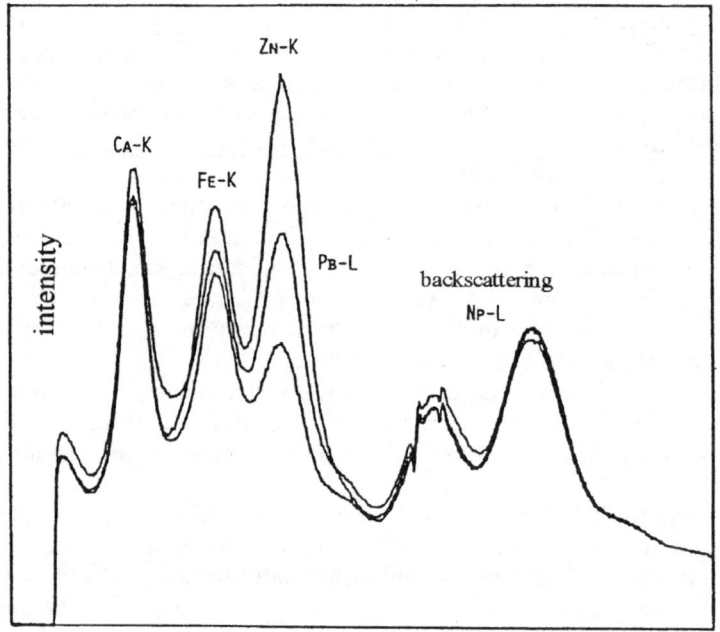

Fig. 4.7. X-ray spectra of samples with different zinc content

tation in several process stages. The products are characterized with regard to their K, Al and Fe contents; in the case of feldspar the K (K_2O) content in particular determines the quality. The task of process control consists of keeping the Al_2O_3 contents in the feldspar and quartz process cycles at a specified level. The on-line analysis of the process flows leads to more suitable products by means of lower contamination levels and to a higher yield as a result of the avoidance of process disturbances and to uniform production of products in accordance with specifications [15].

In the combined pre-processing of *gold* and *uranium* in South Africa on-line XFA methods have also rendered outstanding services [16]. For the determination of baryte contaminations in *fluorspar* concentrates of a flotation process an

Table 4.4. Radionuclides as radiation sources for energy-dispersive XRF

Radionuclide	Half-life period (years)	Energy of the emitted radiation (keV)
^{55}Fe	2.7	5.9 (Mn-K-X-ray)
^{238}Pu	86.4	13–20 (U-L-X-ray)
^{241}Am	433	14–21 (Np-L-X-ray)
		60 (Gamma)
^{244}Cm	18	14–21 (Pu-L-X-ray)
^{109}Cd	1.3	22, 25 (Ag-K-X-ray)

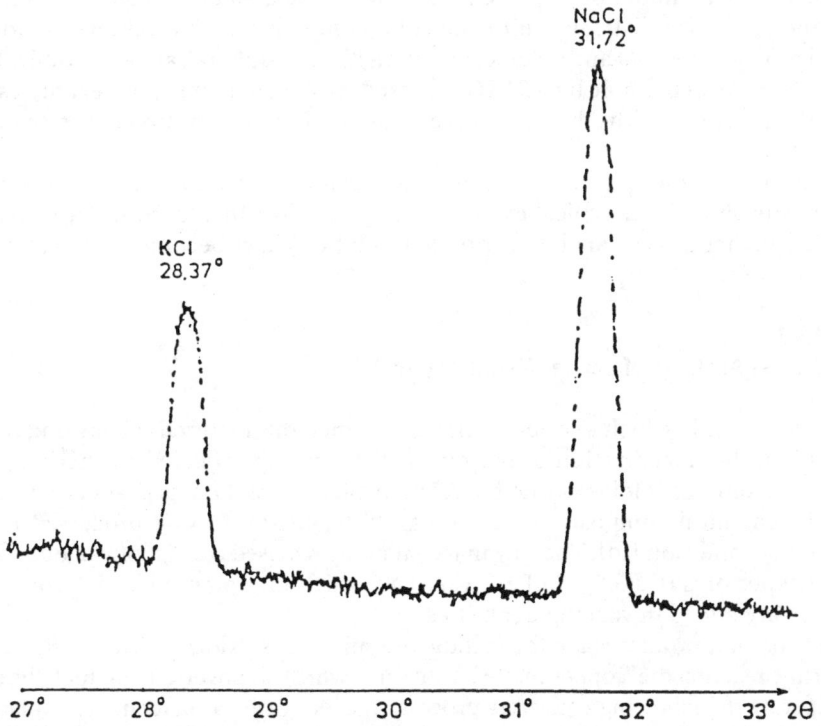

Fig. 4.8. X-ray diffraction spectrum of sylvinite (application of XRD in the case of sylvinite dressing)

on-line method was developed which is independent of density fluctuations of the flow of ore sludge [17].

The energy-dispersive XFA has also proved successful for on-line process control in raw magnesite pre-processing. *Magnesite* is an important raw material for the production of refractory building materials which are used as refractory linings of metallurgical melting vessels, cement and lime kilns. The raw magnesite obtained by mining is treated by flotation, i.e. is freed of its contaminations by a sink/float process. The economical controlling of the dressing process by an on-line control system and thus the optimization of the subsequent sintering process necessitates the on-line analysis of the various products of this process chain [18]. The determination of the Ca (CaO) and of the Fe (Fe_2O_3) is therefore of priority interest for the process control system.

In order to excite the fluorescence radiation, in this case as radiation source for the Ca determination the isotope ^{55}Fe (40 mCi) and for the Fe determination the isotope ^{244}Cm (100 mCi) are used.

The analyzer system removes a part of the ore sludge flows containing solids portions of 10–40% m/m, for example at intervals of 5 min and leads it past the measuring window of the flow cell (Fig. 4.9). The precision of the measurements depends on a number of influencing variables. The greatest influence is exerted by the grain size distribution of the solid components. As a relative standard deviation for the CaO and Fe determination about 5–10% rel. are achievable. The process control possible by XFA is expressed, as shown in the above examples, in the improvement of the dressing success and in a homogenization of the product properties.

The above description cannot make any claims to completeness. It is meant merely to show some typical examples of application. In addition, a large number of further process analytical problem solutions have become known and are conceivable.

4.3.3
Process Analytics of Copper Metallurgy

In those cases in which a copper works has to melt raw materials of varying compositions, homogenization of the raw material used is effected by mixing processes. From this it follows that for economical process control exact knowledge of the chemical composition of the particular mixture to be processed is an essential condition [19]. The raw material mixtures used consist of various sulfidic copper ores ($CuFeS_2$, Cu_5FeS_4, Cu_2S), to which interplant materials and flux agents are added in varying quantities.

In the first process stage the sulfidic ore mixture is oxidized. During this exothermic reaction the "copper matte" is obtained which is converted into metallic raw copper by the following converter process. In order to avoid inadequate oxidation or oxidation going too far by exact setting of the particular O_2 quantity required, continuous determination of the elements Cu, Fe, S and Si (SiO_2) is necessary.

Process analytical control starts with discontinuous, automatic sampling (nine samples per hour) from the conveyor belt which feeds the raw material mixture to the kiln. After dividing the samples the powdery sample is fed to the measuring cell of an energy-dispersive X-ray fluorescence analyzer. The time lag between the time of the sampling and the material entering the kiln is only 30 s and can be neglected. Calibration is carried out with chemically analyzed operating samples which at certain times are removed from the sample flow directly after the measuring cell. The measuring time for the elements Cu, Fe, S, Pb, Zn, As and Ca is 3 min. The analytical results are converted into process parameters by regression models.

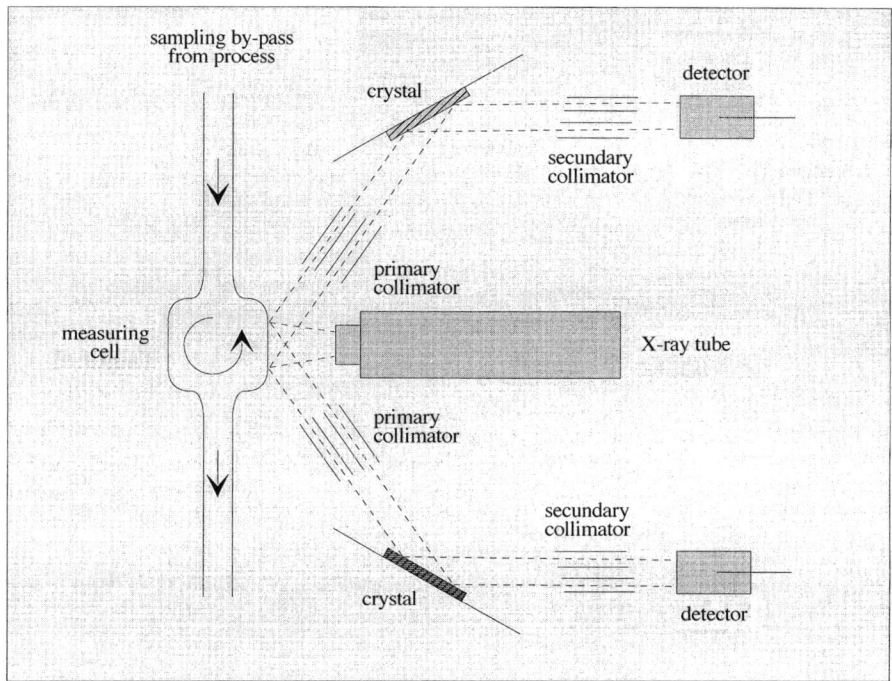

Fig. 4.9. Flow cell of an analyzer system for the control of pre-processing plants

The on-line analysis of the feed-in of raw materials led to a homogenization of the kiln and converter operation, to a better utilization of energy with lower copper losses as well as to a higher yield and to a better copper anode quality.

4.3.4
Continuous Control of Iron Ore Sinter Production

The economical mode of operating a blast furnace is tied to a physically and chemically pre-processed raw material which has a high iron content and has been optimized with regard to its chemical composition (see Sect. 6.2.1).

The operating materials fed into the blast furnace (burden) must have, among other things, a certain grain size in order to make possible trouble-free and economical operation. As the iron ores in the state as delivered rarely meet the requirements of the blast furnaces, they are reduced in size to about 5–6 mm in special crushers and screening units and after that screened to separate excessively coarse (oversized) grains and the fine portions (undersized grains). The fine portions pass together with various metal works by-products rich in iron and the "fine ores" traded in large quantities on the world market – i.e. often ores which have been concentrated by means of flotation – into the sintering and pelletizing plants (Fig. 4.10) where they are converted into materials suitable for use

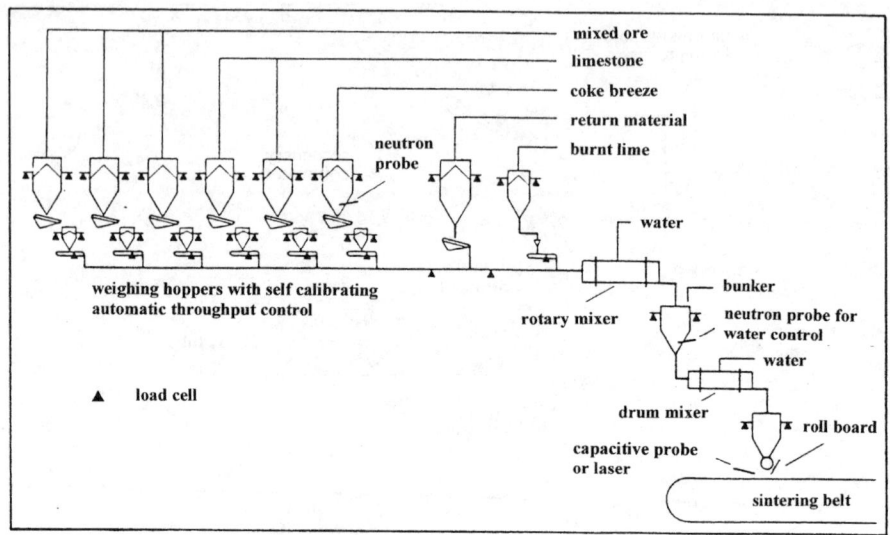

Fig. 4.10. Mixed material blending in the case of iron ore sinter production (by permission of Verlag Stahleisen GmbH, Düsseldorf)

Fig. 4.11.
Material flow of the sintering process

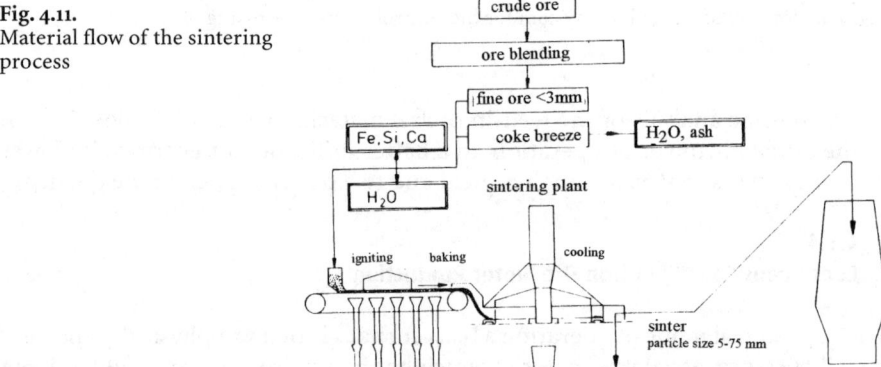

in the blast furnaces. The "sinter" nowadays represents the preferred material for use in the blast furnace, adjustable to the particular mode of operation.

In the case of the sintering process – the flow of material is shown in Fig. 4.11 – the material to be sintered (grain size <3 mm), possibly with the addition of lime or dolomite (to adjust the desired $CaO+MgO/SiO_2$ ratio!), is mixed with coke breeze (5–6%) as a supplier of energy and heated to such a temperature that the individual ore grains soften on their surface and fuse with each other. The

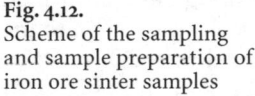

Fig. 4.12.
Scheme of the sampling
and sample preparation of
iron ore sinter samples

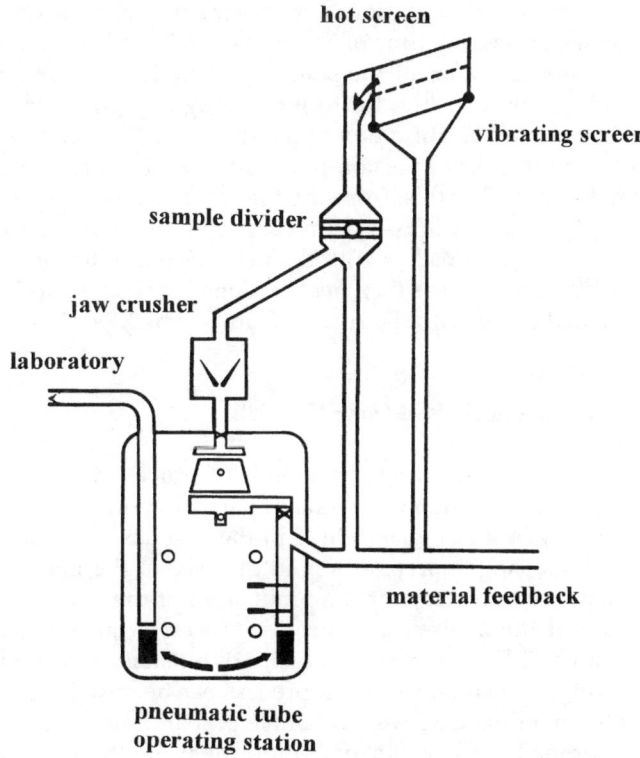

hard, pore-rich sinter is obtained which withstands the high mechanical stresses and because of its gas permeability makes easier the penetration of the reduction gases flowing towards it in the blast furnace.

With the control of sinter production the process analytical tasks on the way from pig iron to steel production and processing start. While the iron ores used are generally analyzed discontinuously using well-known methods on the basis of samples representing large stock quantities, optimum control of the process requires, among other things, continuous control of the product quality, which at the same time means a contribution to the automation of this process [20]. As an analytical principle wavelength-dispersive X-ray fluorescence spectrometry is used.

After the sintering belt sampling takes place continuously and, of the roughly 700 kg/h of material obtained, the 6–12 mm portion is separated out by screening. This portion is reduced to about 5 kg/h which is then crushed to below 3 mm and homogenized (Fig. 4.12). At 30 min intervals, of this collective sample a laboratory sample of about 400 g is fed with a pneumatic tube system to the sample preparation device. Here, the fine grinding of the sample with the addition of grinding and pressing aids takes place, followed by pressing into a tablet which is then automatically fed to the spectrometer. Total sample preparation including the cleaning of the vessels takes about 14 min.

Double determinations of the elements Mg, Al, Si, Ca, Ti and Fe are performed with a measuring time of 20 s and the results are automatically passed on to the process computer of the sintering plant. In a process model they can serve as influencing variables for feed belt weighing for limestone, sand and olivine. At 6 h intervals recalibration of the spectrometer is effected automatically.

As a further important process for ore pre-processing, pelletizing should be mentioned. For this, from the fine ores (grain size<0.05 mm) small ore pellets with a diameter of 15–35 mm are produced. The initially moistened ore mixture is formed into pellets with the help of rotary drums or rotating plates ("green pellets"), which are then heat hardened at 1000 –1100 °C. Process control can be carried out as with the sintering process by XFA.

4.3.5
On-Line Analysis of Lump Materials

In the above-mentioned methods of process analytics the solid matters were removed from the process flows partly in powder form, as a slurry or discontinuously in lump form. While in the first cases mentioned analysis takes place without sample preparation, in the case of the lump material various process steps are necessary for the production of the analysis sample. However, in the case of the analytical control of process sequence the question of the direct testing of lump materials which are constantly fed into a process, or as a final product constantly leave the process, can be raised. Here, as in some of the examples mentioned above, no sample preparation is possible as the material to be examined is in permanent motion, for example on a conveyor belt. For the explanation of this process analytical question the following examples may serve in which various analytical principles are used.

One possibility is provided by XFA. At a height of 15–30 mm above the material to be analyzed an X-ray tube and corresponding detectors (proportional counters), for example, are positioned (Fig. 4.13) and in measuring steps of 0.2 s, 300 measurements per minute are made, the mean values of which are determined over a definite period. There are versions with which moved material up to a lump size of 100 mm can be investigated continuously and simultaneously for two elements [21]. As an example of this type, the determination of Ca and Fe during *limestone open-cut mining* can be mentioned. Calibration takes place with the help of a regression calculation in which the pulse rates and the chemical analysis of calibration samples are included.

During the continuous monitoring of the *water content of bulk materials*, e.g. of hard coal coke or sinter blends, measurement of the neutron scattering is used. As a supplement to the X-ray fluorescence spectrometric monitoring of sinter production regarding a constant product quality, the determination of the water content of coke breeze serving as the supplier of energy and finally of the water content of the sinter raw mixture with the help of radionuclides represents a considerable contribution to the optimization of the process (low energy consumption, uniform mode of operation, high product quality) [20]. Figure 4.10 shows schematically the mixed product pre-processing between the ore stocks

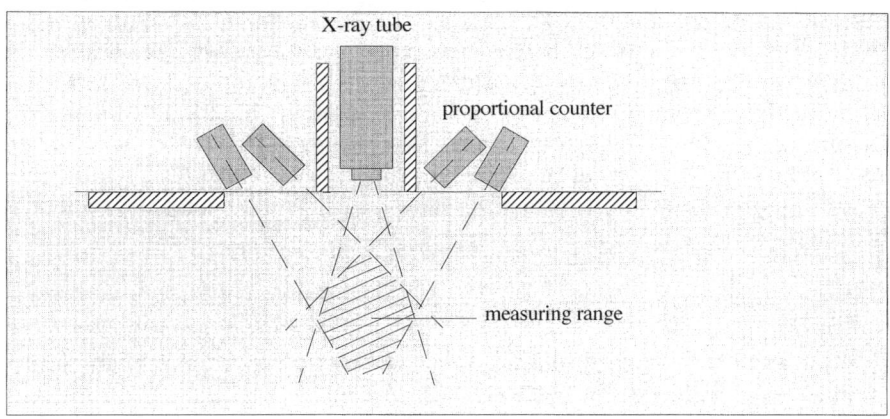

Fig. 4.13. Measuring principle for X-ray fluorescence spectrometric analysis on conveyor belts

("mixing beds") and feeding onto the sinter belt (see also Fig. 4.11). As a function of the sinter belt speed and the filling levels of the feed and intermediate bunkers, the mixed product flow rate is controlled. The water content of the coke breeze continuously measured with neutron probes makes possible the addition of the coke breeze corresponding to the demand for dry fuel. The control of the water content after mixing and the moistening of the product is also carried out with the help of a neutron probe.

The calibration of such measuring equipment is performed with samples which are taken from current production, the water content of which has been gravimetrically determined. Thus, the composition and the bulk density of the product and the geometry of the measuring cell have to be taken into account. The precision of the method (2 s) amounts to $\pm 0.5\%$ m/m H_2O.

Of great influence are fluctuations of the hydrogen percentage in the coke which results from different coking conditions. Here contents of 0.4–0.8% m/m are observed, this hydrogen portion referring to all protons which are not combined to oxygen. According to the method described here for coke breeze and raw sinter mixtures, a continuous determination of the water content in the blast furnace coke can also be effected as a contribution to the controlling of the input of fuel during the blast furnace process [22].

The last example comes from hard coal blending. The coal obtained by mining shows different inerts and is therefore not directly useable industrially. It requires blending in order to reduce the undesirable inert portion (ash) to a tolerable level and to bring about a homogenization of the raw material properties, and thus of the ash content. For control of this dressing process continuous determination of the ash content in the product flows of the different process stages is a necessary condition. For this only contactless measurement which should be independent of the conveyor belt loading and the bulk density of the product can be considered. The problem is also solved by the use of radionuclides. The radiation emitted by a gamma radiation emitter is weakened when being transmitted through matter. The ash constituents of the coal have a higher ordinal number than the coal substance and thus absorb more strongly. Thereby, the absorption becomes a direct measure of the ash percentage. A scintillation counter which is highly sensitive to gamma radiation measures the remaining residual radiation and supplies a signal proportional to the radiation intensity. By the selection of the radiation sources (e.g. ^{241}Am/^{137}Cs) and of the measuring procedure, measurements independent of bulk height and bulk density are possible [23].

4.4
Suppliers for Sampling and Sample Preparation Devices and for Analytical and Automation Systems

4.4.1
Sampling and Sample Preparation

Friedrich Debus Maschinen- und Apparatebau, Industriestrasse7, D-52134 Herzogenrath
Maschinenfabrik Herzog GmbH & Co, Auf dem Gehren 1, D-49086 Osnabrück
Krupp Polysius AG, Graf-Galen-Strasse 17, D-59269 Beckum
F. Kurt Retsch GmbH & Co KG, Rheinische Strasse 36, D-42781 Haan
Siebtechnik GmbH, Platanenallee 46, D-45478 Mülheim/Ruhr
SICON GmbH, Flurstrasse24, D-42781 Haan

4.4.2
Wavelength-Dispersive X-Ray Fluorescence Spectrometers

Fisons Instruments Vertriebs-GmbH, Peter-Sander- Strasse 43, D-55252 Mainz
Oxford Instruments Deutschland GmbH, Kreuzberger Ring 38, D-65205 Wiesbaden
Philips Industrial Electronics Deutschland GmbH, Miramstrasse87, D-34123 Kassel
Rigaku Europe GmbH, Monschauer Strasse 7, D-40549 Düsseldorf
Siemens AG, Röntgenanalytik, Postfach, D-76181 Karlsruhe

4.4.3
Energy-Dispersive X-Ray Fluorescence Spectrometers

ASOMA Instruments Vertriebs-GmbH, Ingolstädter Strasse 68 a, D-80939 München
Österreichisches Forschungszentrum Seibersdorf GmbH, A-2444 Seibersdorf
Outokumpu KM-Analytik GmbH, Königsteiner Strasse 98, D-65812 Bad Soden
Oxford Instruments (see above)
Spectro Analytical Instruments GmbH, Boschstrasse10, D-47533 Kleve

4.4.4
X-Ray Diffractometers

Fisons Instruments (see above)
Outokumpu KM-Analytik (see above)
Philips Industrial Electronics (see above)
Rigaku (see above)
Siemens AG (see above)

4.4.5
Radiometric Analytical Systems

Laboratorium Prof. Dr. Berthold, Calmbacher Strasse 22, D-75323 Bad Wildbad

4.4.6
Remarks

The preceding compilation cannot make any claim to completeness. For further information reference should be made to the literature and the market surveys published in "Nachrichten für Chemie, Technik und Laboratorium", the news of the Gesellschaft Deutscher Chemiker (GDCh) (Society of the German Chemists). Further information can be obtained from the References to Chaps. 3 and 6.

References

1. Kraft G (1980) In: Probenahme – Theorie und Praxis – Schriftenreihe der GDMB Gesellschaft Deutscher Metallhütten- und Bergleute, Heft 36, Verlag Chemie, Weinheim, S. 1
2. DIN 51 701: Prüfung fester Brennstoffe – Probenahme und Probenvorbereitung Tl. 2: Durchführung der Probenahme; Tl. 4 und Beiblatt: Geräte; Bergbau-Betriebsblatt BB 23017 – Probenahme von Kraftwerkskohle; Empfehlung für Bau und Betrieb automatischer Einrichtungen – Bergbau-Betriebsblatt BB 23018 – Probenahme von Kokskohle; Empfehlungen für Bau und Betrieb automatischer Einrichtungen
3. Koch W, Roeder KP, Dobner W, Huber G (1970) Thyssenforsch. 2:6
4. Wende V (1989) In: Schriftenreihe der GDMB Gesellschaft Deutscher Metallhütten- und Bergleute, Heft 55, S. 127
5. Ehrhardt E et al. Röntgenfluoreszenzanalyse – Anwendung in Betriebslaboratorien, VEB Deutscher Verlag für Grundstoffindustrie, Leipzig
6. Jenkins R (1994) In: Ullmann's encyclopedia of industrial chemistry, vol B5. VCH Verlagsgesellschaft, Weinheim, p 675
7. Staats G (1989) Fresenius' Z Anal Chem 334:326
8. Neff H (1962) Grundlagen und Anwendung der Röntgen-Feinstruktur-Analyse, 2. Aufl., R. Oldenbourg, München
9. Gieren A, Paulus E F (1994) In: Ullmann's encyclopedia of industrial chemistry, vol B5, VCH Verlagsgesellschaft, Weinheim
10. Uhlig S, Burgäzy F (1993) Kontrolle, November:14
11. Trost A (1965) Z Instrumentenkde 73:329
12. Oesterle G (1995) Prozeßanalytik – Grundlagen und Praxis. Oldenbourg, Munich
13. Donhoffer D (1984) Berg- u. Hüttenmän. Monatsh. 129:453
14. Saarhelo K On-line X-ray analysis in industrial applications. Firmenschrift Ref. 3881 481–4 der Fa. Outokumpu Electronics Oy, SF-02201 Espoo

15. Saarhelo K, Paakkinen U, Pennanen P (1988) On-line analysis in industrial mineral applications. 8th Industrial Minerals Inst. Congress, Boston
16. de Jesus ASM, Basson JK (1977) Int J Appl Radiat Isot 28:873
17. Lubecki A, Wiese K, Winkler K (1980) Kernforschungszentrum Karlsruhe, KfK-Report 2936:2
18. Weiss V (1991) Radex-Rundsch H 1/2:400
19. Davenport WG, Partelpoeg EH (1987) Flash smelting: analysis, control and optimization. Pergamon, Oxford
20. Buckel M, Kersting K, Kister H, Lüngen H-B (1990) Stahl u. Eisen 110, Nr 2:43
21. Beltcon 200 Firmenschrift Ref. 3882 240–4 He Ed. 2 der Fa. Outokumpu Electronics Oy, SF-02201 Espoo
22. Neuhaus H, Hombeck F, Kühn W (1962) Stahl u. Eisen 82:1017
23. Asche-Monitor-System LB 420 Firmenschrift der Fa. Laboratorium Prof. Dr. Berthold, Wildbad

5 Process Analytics in the Chemical Industry

5.1
The Role of Analytics in the Chemical Industry

The role of analytics has become of increasing importance in all economic fields and therefore also in the chemical industry according to changing needs and market requirements and is subjected to continuous change (see Sect. 1.3). The methodical and instrumental developments enable today a closed monitoring of process chains and lead to the inclusion of analytics in the entire development, production and marketing course of a company.

As in other industrial branches, analytical chemistry has a long tradition in the chemical industry. The development led to an organization (see also Sect. 1.4) by which the analytics has become a competent partner for the optimization and control of the process chain "production" and for the support of research activities, product development and commercial exploitation [1]. The requirements for rapid, reliable, reproducible and true on-site-analyses as presupposition for a closed loop control of production processes cannot be fulfilled in most cases by a centrally located, method-oriented analytical laboratory, but demands today are for process-oriented, low cost analysis systems, that can be operated on-line, at-line or even in-line [2]. A high degree of automation, easy operability and low maintenance expenditure, respectively, are necessary prerequisites. Analytical principles which are applied to an increasing extent for these new tasks are gas chromatography [3], VIS [4] and NIR spectrometry [5]. In particular, the latter principle serves in recent times to perform fast identity and purity examinations [6]. By the application of glass fibre optics, measurements can possibly also be made in-line, i.e. directly in the process medium [7].

The use of chemical sensors in process analytics (see Sects. 2.3.3 and 3.2.1.2) is still at the beginning of a tempestuous development [8]. A remarkable number of chemical sensors for numerous analytical tasks are already available today [9–11].

For routine analyses, especially in the framework of process analytics, which are only of significance within a company the nowadays much discussed "outsourcing" should not or only exceptionally be considered. Each analysis externally produced must be checked again by the ordering company for plausibility, trueness and "useability". Furthermore, problems arise regarding the responsibility when utilizing the purchased services and over the protection of confidentiality; queries cannot be answered directly, frequently incompetently and often late, so that the presumed economic advantage turns into quite the opposite.

In Chaps. 2–4, the methods of process analytics of gaseous, liquid and solid media have already been explained, as they are also used with process courses in the chemical industry. The following two specific examples as supplements to these descriptions, from the multiplicity of the industrial chemical processes, are of importance: they are the control of dye-stuff production as a typical example of a discontinuous regulated process (batch process) and the process analytics in biotechnology; both classical and modern, forward-looking chemical process technology are involved.

5.2
Process Analytics for Dye-Stuff Production (Batch Process Control)

The procedures for the production of dye-stuffs are discontinuous led chemical processes, by which, for example, an aromatic amine is converted by continuous charge of acid chloride. The primary product formed has to be processed by different preparation steps to the final product. In order to receive in this case a qualitatively faultless final product, the amount of amine may not exceed 100 mg/l in the reaction mass.

For the control of this batch process the "flow injection analysis" (FIA), which has also been applied successfully to other tasks (see Sect. 3.2.1.1), has been proved as suitable [12]. As the concentration of educts and products change during a batch process over several orders of magnitude, FIA systems are required that show a high dynamic measuring range together with a precision as high as possible. The measuring range and the accuracy of a FIA system are narrowly connected with the mode of detection and the evaluation of the FIA measuring signals. By combination of photometric detection with gradient techniques, FIA systems are obtained which possess a measuring range of 5 and more decades as well as a precision sufficient for process control (1–2% rel. standard deviation) [12]. Here, several procedures are possible, of which as an example the peak height measurement combined with gradient methods (peak width measurement) is selected.

For control of a batch process, a less precise determination is in many cases sufficient at the beginning of the reaction if the concentration of the process material is high and therefore a great dilution of the sample is necessary. At the end point of the process the desired concentration or the concentration to be reached must be determined very exactly in order to reach an optimal process operation. In this case, a FIA system can be elected so that the range of the end point concentration over the peak height can be measured linearly. When injecting samples of higher concentrations, a chemical or optical saturation occurs in the system with the consequence that the peak height hardly increases with increasing concentration of the analyte but remains nearly constant. However, the peaks become wider, and in the descending flank of the gradient, which cor-

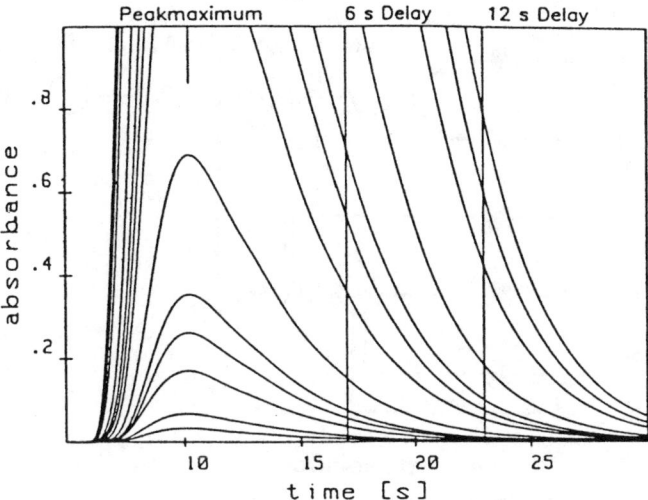

Fig. 5.1. Principle of peak height measurement combined with a gradient evaluation method (peak width measurement) (by permission of Springer Verlag GmbH. & Co KG, Heidelberg)

responds to the signal of the increasingly diluted sample, the concentration range linearly measurable by the FIA system is run through. A simple evaluation method now consists of measuring the peak maximum and several defined positions in the descending flank of the gradient (see Fig. 5.1 for excerpts from the graphic diagram of those FIA signals obtained from calibration solutions which cover a concentration range of four orders of magnitude). Each of these temporally defined measuring points in the gradient corresponds to a restricted concentration range with linear calibration. The measuring range of a FIA system with this mode of evaluation is only limited by the precision, increasingly becoming more difficult with high dilutions [12].

In the case of the example shown in Fig. 5.1 [12] the peak maximum and the extinction values of the intersections of the signals with the verticals 6 and 12 s after the peak maximum that corresponds approximately to the signal maximum of a sample diluted by the factor 5 or 40, respectively, are measured. The calibration curves are obtained by plotting the measured extinction values as functions of the concentrations. The precision of the determination at the peak maximum is stated with <0.5% rel. standard deviation. It decreases in the gradient with increasing distance from the peak maximum and amounts to approximate 1–2% rel. standard deviation at 12 s after the peak maximum [12].

FIA systems of the type mentioned allow dilutions up to a factor of 100,000, so that additional sample preparation equipment is not necessary. In the analyzer used here (Fig. 5.2) the amine is coupled with diazonium salt and the azo dye formed is measured photometrically. The unstable diazonium ion is generated continuously from acidified 1,4-nitroaniline and nitrite. Excessive nitrite is disintegrated by amidosulfuric acid. The sample taken out of the reaction vessel is

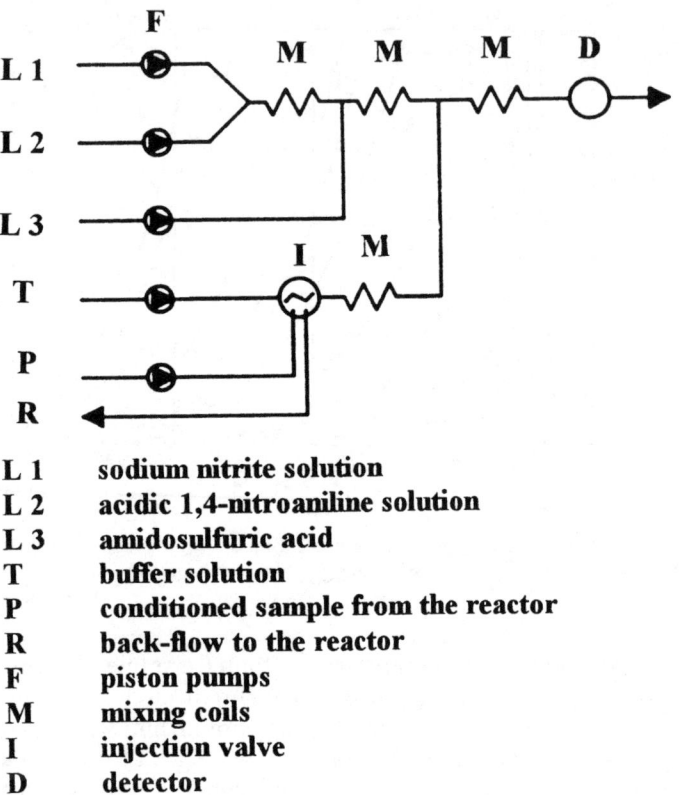

L 1	sodium nitrite solution
L 2	acidic 1,4-nitroaniline solution
L 3	amidosulfuric acid
T	buffer solution
P	conditioned sample from the reactor
R	back-flow to the reactor
F	piston pumps
M	mixing coils
I	injection valve
D	detector

Fig. 5.2. Scheme of a monitor for the continuous determination of amine

pH conditioned in a buffer and brought together with the reagent [13]. One analysis cycle lasts 2 min, i.e. the input time of the controlled system: acid chloride addition – mixing in the reactor – sampling – analysis lies in the range of minutes. The calibration is performed in the usual manner with graded calibration solutions.

5.3
On-Line Analytics and Biosensorics in Biotechnology

5.3.1
Biotechnology and Bioprocesses

Biotechnology confronts us first of all with regard to fermentation processes, fat and protein production as well as sugar and starch production as classically appearing branches of chemical process technology, although production of vitamins, hormones and enzymes for most different fields of application also belongs in this historic picture. However, recently biotechnology has undergone a development that has opened up completely new ways resulting from work of numerous researcher groups at universities and other research institutions at home and abroad. Today, there is no doubt that biotechnology possesses an immense importance for the future of mankind.

The biotechnology of amino acids, vitamins, enzymes, antibiotics, alkaloids, pharmaceuticals and therapeutics (cytostatics) support the preceding opinion. Another noteworthy field is the energy production from biomass, where, however, the application of vegetable oils in the fuel sector ("biofuels") as an ecologically significant contribution is still seen as controversial. In this field, the development is directed to vegetable oils for industrial use modified by genetic technology.

The latest direction of research in this discipline, characterized by rapid developments, is called "evolutive biotechnology" which aims at the adaptation to product requirements by molecule syntheses and the finding of new active components combined with a decrease in the time for development and an optimization of the specific effect of pharmaceuticals and diagnostics. Another direction of development attempts the "construction" of not natively found enzymes for the catalysis of reactions which would normally only be chemically accessible.

5.3.2
Total Chemical Analysis Systems (TAS)

The modernity of technological development corresponds to process coinciding analytics. On-line process analytics and biosensorics [14, 15] create the necessary prerequisite for the successful investigation of biotechnical processes and their transfer to the industrial scale. Computer-aided complete analysis systems are required (TAS = total chemical analysis system [16]) that allow a rapid and selective determination of low concentrations of substances in the case of biosensorics by combining chemical, biological, optical and electronic components [17]. For the continuous control of bioprocesses by means of biosensors there are different possibilities (Fig. 5.3) which depend on properties of the process materials and possibly generated by-products as well as on the construction of the bioreactors.

From the literature [18, 19] it can be quoted that carbohydrates are the most frequently named analytes, followed by ethanol, lactates, amino acids, acetates and diacetyl. The analytic development is definitely determined by the trend to multicomponent analysis. The widespread industrial application of biosensors still stands at the very beginning for different reasons [15, 20]. One problem is that the analytical tasks posed in each case cannot generally be solved but only processed specifically. But nevertheless it can be assumed that biosensorics will gain increasing importance for the optimization of processes and for the improvement of product quality [21, 22].

Fig. 5.3.
Principles for in-line and on-line control
of bioprocesses

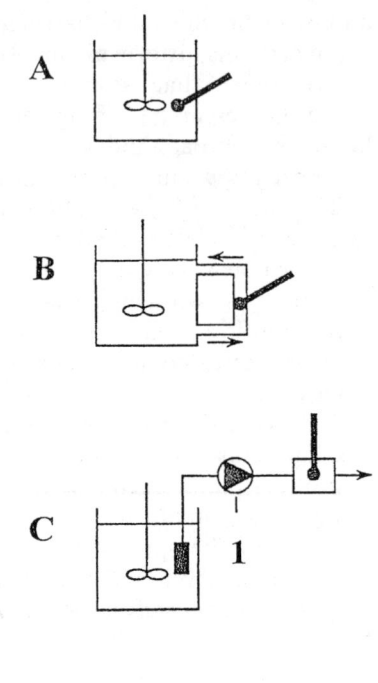

A = in-line (in situ-) measurement
B = in-line-measurement in a bypass
C = on-line-measurement by means of a ex
    situ-sensor
D = bypass sampling and on-line measurement
    by means of a ex situ sensor
1 = pump
2 = filtration unit

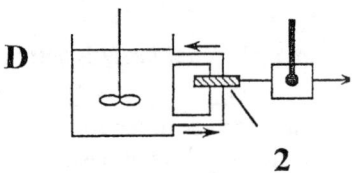

The use of most of the chemical and biosensors described in the literature has been restricted until now to samples with a simple chemical matrix [23]. Direct measurements in complex media are in general (still) not possible because of the disadvantageous influence on the reliability and ability for unequivocal statements. In order to avoid those factors restricting the applicability, the above-mentioned, more flexible applicable total chemical analysis systems (TAS) have been proposed [24], where analytic principles like high performance liquid chromatography (HPLC) [25], capillary electrophoresis [26], and flow injection analysis (FIA) [27] form the basis of the method. Even the last mentioned method [28] (see also Sect. 5.3.3) offers the possibility to adapt the sample portion to the selectivity and the dynamic range of the sensor used. By chemical reaction of the analyte or extraction of the disturbing accompanying substances the desired selectivity of the method can be achieved, while an adaptation to the dynamic range of the sensor is reached by enrichment or dilution steps over several orders of magnitude. Furthermore, the lifetime of the sensor in an FIA system is extended considerably as its contact with the sample only lasts for some seconds. The frequently required miniaturization of such systems has played up to now a subordinate role because of the considerable difficulties with the manufacture of the necessary piping of very small volumes, and corresponding pumps, and detectors [29]. Lately there have been suggestions for the construction of FIA systems on the microscale which open completely new applications for this technology [27].

5.3.3
Flow Injection Analysis (FIA)

In biotechnology, fermentation processes always play an important role. In these cases the growth and the production rate of a microorganism population depend directly on its environment. Accordingly, the growth-decisive parameters and the substrates must be controlled during the fermentation for the optimal operating of a fermentor. This aim can only be reached by the on-line measurement of these parameters during the whole process course. Here, one problem among others is that the samples have to be taken out of the reactor under sterile conditions and must subsequently be filtered, diluted and prepared for analysis. As until now the sequential off-line determination of the most important chemical parameters was the usual practice, the flow injection analysis (FIA) now offers a promising solution for the on-line control of industrial fermentation processes. By inclusion of sample dilution and conditioning steps between sampling and analysis, this method can be applied to the substrate analysis over the total process concentration range [30].

An automatic analysis system of this type consists of a sterilizable microfilter, a dosing unit for samples and reference materials, a degassing and dilution unit and the FIA piping with a UV/VIS spectrophotometric detector which allows, for example, the determination of glucose (peroxidase), ethanol (peroxidase), ammonia (indophenol blue method) and phosphate (molybdenum blue method) [30]. The total cycle time amounts in this case to 90 s for sample preparation and analysis of one sample, i.e. the attainable sample frequency is 30 samples per hour. The relative standard deviation under the conditions described is stated to be approximately 1.1–1.6%. It has been proven on the basis of these data and numerous investigations that the FIA method allows one to predict reliably the end point of a batch fermentation process.

The photolithographic manufacture of microflow systems enables the miniaturization of the FIA [31] and opens up new fields of application for this analytical technique. A set of planar elements of the same diameter with different borings and cavities which are stacked one on top of the other and are connected to a module replaces the classically valve regulated FIA piping [27]. The symmetry of the elements, which, for example, can possess a thickness of 2 mm and a diameter of 50 mm, allows one to produce different connections between different piping tracks (see Fig. 5.4). The insertion of this module into an analysis system, which serves for the input of the sample, mixing of the reactants and possibly for dilution, is shown schematically in Fig. 5.5. Materials for such components include plastics, metals and ceramic materials [32].

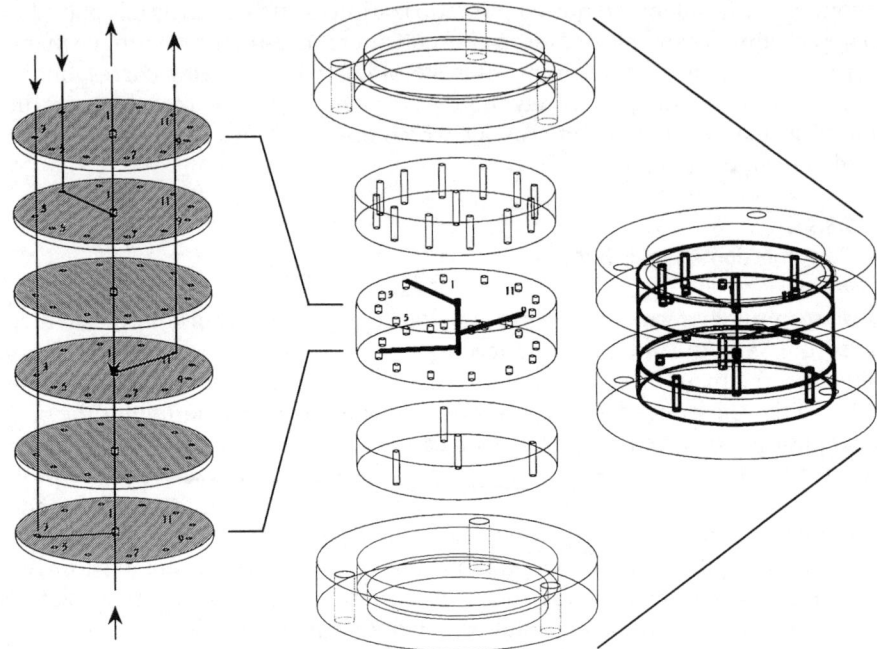

Left: Exploded view of a vertically aligned and stacked set of elements performing the functions of three modules.
Middle: Stack aligned and compressed with upper and lower caps shown along the holder.
Right: Stacked series of modules ready for use (tightening bolts not shown).

Fig. 5.4. Modules for microflow systems [27]. *(By permission of Elsevier Science B.V., Amsterdam)*

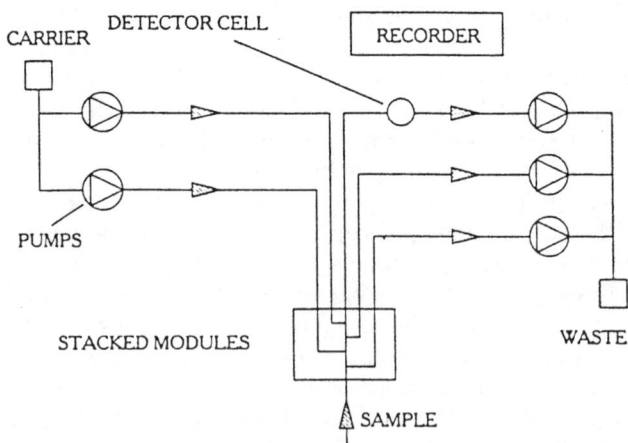

Fig. 5.5. Insertion of a microflow system into an FIA system (schematically) (by permission of Elsevier Science B.V., Amsterdam)

Further developments of a TAS in microscale (μ-TAS) [33] aim at a size of microflow module of less than 1 cm³. Further research on these valveless flow systems concerns the integration of the detectors and pumps. It is expected that by these analysis systems limits of detection of 10^{-5} mol/l analyzing sample volumes of less than 100 nl can be reached. A μ-(micro-) TAS that delivers the desired analytical information within a few seconds until minutes is similar to a chemical sensor (see Fig. 5.6) [34]. The theoretical bases of this technology and other concepts suitable for a TAS or a μ-TAS have already been described in detail [31, 34, 35]. In comparison to chemical sensors, miniaturized total chemical analysis systems cause less problems regarding, for example, the selectivity, since here chemical procedures proved for sample preparation can be included. The integration of separation methods also enables multicomponent monitoring with a single measuring arrangement. Although the last-named systems are still in the research and development stage, they are nevertheless briefly mentioned since they will feature decisively in analytics of the future.

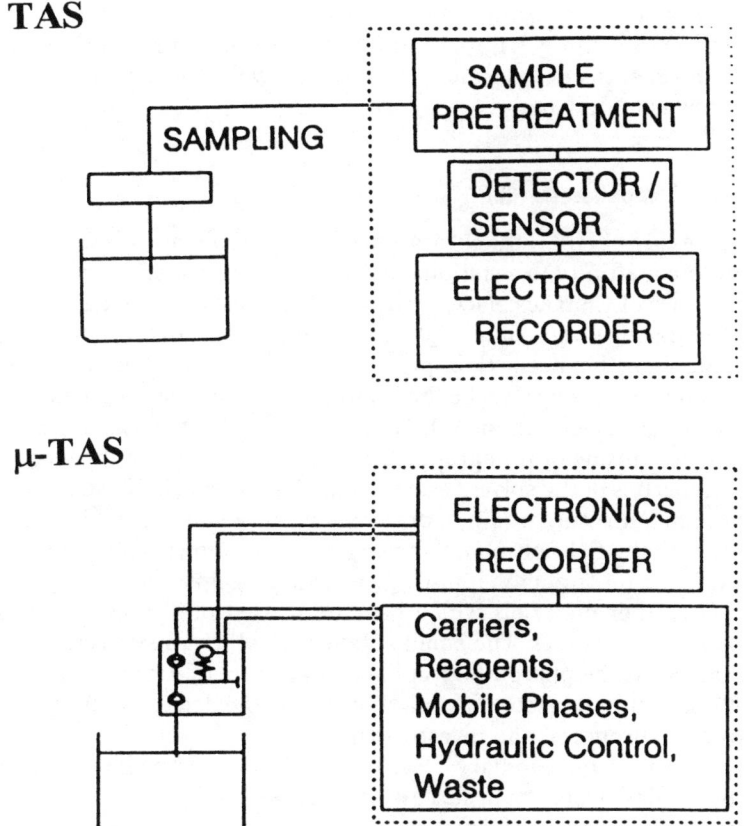

Fig. 5.6. Schematic diagram of a total chemical analysis system (TAS) in comparison to a miniaturized TAS (μ-TAS)

5.3.4
High Performance Liquid Chromatography (HPLC)

One way for the development of total chemical analysis systems (TAS) is as already mentioned in the application of flow injection analysis in connection with separation methods like high performance liquid chromatography (HPLC) or capillary electrophoresis (CE) [36] which have found widespread application in organic analytics for many years. A prerequisite for the application of HPLC in a TAS or μ-TAS consisted in the development of "microcolumns" with an inner diameter of few microns up to 1 mm that allow analysis of samples of very small volumes in a short time.

A concept for HPLC on the microscale consists of the manufacture of a silicon chip in the size of $4.5 \times 25 \times 0.75$ mm by photolithographic techniques and etching processes which contain, besides the sample input, the separation column and an optical cell for the detector [25]. The separation column possesses a volume of 0.5 μl, and the stagnant volumes of the piping is below 2.5 nl. One advantage in the application of an HPLC chip in a TAS compared to CE is the easy transferability and application of HPLC standard methods. It can be assumed that these micromethods will play an important role in the course of the next years for the analysis of extremely small sample volumes for on-line measurement in total chemical analysis systems.

5.3.5
Capillary Electrophoresis (CE)

While in the case of HPLC the separation of the components of a sample is based on the different affinity to a stationary phase (e.g. in a column) [37], the separation in the case of CE takes place by the variable effect of an electric field on the single components of the sample [38]. By etching procedures a TAS-corresponding CE module with a network of piping on the metre and millimetre scale can also be produced which can take the sample input, the sample preparation, and the electrophoretic separation unit. The selection of the material depends on the workability by means of lithographic procedures, the type of the electric field, and the properties of the solvent and the sample. An especially interesting material for the manufacture of a μ-TAS is silicon, since considerable experience is available from the field of manufacturing electronic components. Since silicon is, however, a semiconductor, it must be insulated against the liquid medium by a dielectric. Other materials which have been used until now are glass, fluoropolymers and silicones. The small volumes in which the detection has to be performed needs the photolithographic manufacture of detector cells with volumes of less than 1 nl up to a few microliters. For the detection different principles have been proposed and investigated [36].

A μ-TAS based on capillary electrophoresis, the elements of which were etched into a glass plate, could successfully be applied, for example, to the analysis of oligonucleotides [39]. By means of a planar silicon structure of size 6×3 cm, on which all items needed for CE have been arranged, the perfect separation and analysis of a mixture of amino acids have been successful [40].

The advantages offered by these analysis systems as fast and selective analyzers with lowest solvent consumption and low yield of waste material recommend them for on-line process analytics and will provide them with great opportunities on the market. Furthermore, the presumed low price per analysis unit will allow the use of these systems at numerous positions in a production plant and can thereby offer extensive process analytics. The improvements to be expected regarding the lifetime and the selectivity of these "chip laboratories" will make them severe competitors of chemical sensors in the future.

5.4
Suppliers for Sensors and Appliances

ABC Biotechnologie GmbH, Boschstrasse6, D-82178 Puchheim
Analytik Jena GmbH, Am Kieshügel 31, D-07745 Jena
ANASYSCON Instrumentelle Analysentechnik GmbH, Vahrenwalder Strasse 7, D-30165 Hannover
Biometra biomedizinische Analytik GmbH, Rudolf-Wissell-Strasse 30, D-37079 Göttingen
B. Braun Melsungen AG, Carl-Braun-Strasse 1, D-34212 Melsungen
Colora Meßtechnik GmbH, Barbarossa-Strasse 3, D-73547 Lorch
Eppendorf-Netheler-Hinz GmbH, D-22331 Hamburg
Gonotec GmbH, Eisenacher Strasse 56, D-10823 Berlin
Ismatic Laboratoriumstechnik GmbH, Vier-Morgen-Strasse 23, D-97877 Wertheim-Mondfeld
Molecular Devices GmbH, Bahnhofstrasse110, D-82166 Gräfeling
Perstorp Analytical GmbH, P.O. Box 1370, D-63085 Rodgau
Pharmacia Biotech Europe GmbH, Munzinger Strasse 9, D-79111 Freiburg
Prüfgeräte-Werk Medingen GmbH, Leßkestrasse10, D-01705 Freital
Dr. Berthold G. Schlag, Am Mühlenberg 19, D-51465 Bergisch-Gladbach

5.4.1
Remarks

The preceding compilation cannot make any claim to completeness. For further information reference should be made to the literature and the market surveys published in "Nachrichten für Chemie, Technik und Laboratorium", the news of the Gesellschaft Deutscher Chemiker (GDCh) (Society of the German Chemists). Further information can be found in the References to Chaps. 2, 3, 4 and 6.

References

1. Fuchs-Pohl G, Härtner H (1995) Nachr Chem Tech Lab 43(1):36
2. Hoyer G-A (1995) Nachr Chem Tech Lab 43(1):242
3. Engewald W (1994) Nachr Chem Tech Lab 42:356
4. Bacher W, Mohr J, Ruprecht R, Schomburg WK, Stark W (1993) GIT Fachz Lab 12:1081

5. Molt K, Ihlbrock D (1994) Fresenius' J Anal Chem 348:523
6. McClure WF (1994) Anal Chem 66:43A
7. Molt K (1994) Nachr Chem Tech Lab 42:M1
8. Carey WP (1994) TrAC 13:210
9. Gauglitz G (1994) Fresenius' J Anal Chem 349:337
10. Reichert J (1994) KfK-Nachr 26:28
11. Bürck J (1994) KfK-Nachr 26:34
12. Thommen C, Garn M, Gisin M (1988) Fresenius' J Anal Chem 329:678
13. Gisin M, Thommen C (1987) Anal Chim Acta 190:165
14. Cammann K, Ross B, Hasse B, Dumschat C, Katerkamp A, Reinbold J, Steinhage G, Gründig B, Renneberg R, Buschmann N (1994) In: Ullmann's encyclopedia of industrial chemistry, vol. B6. VCH, Weinheim, p 121
15. Anders K-D, Plötz F, Scheper T (1991) GIT Fachz Lab 10/91:1093
16. Widmer HM (1983) Trends Anal Chem 2:8
17. Anders K-D, Wehnert G, Thordsen O, Scheper T, Rehr B, Sahm H (1993) Sensors and Actuators B 11:395
18. Schmidt RD, Scheller F (eds) (1990) Biosensors – applications in medicine, environmental protection and process control. GBF Monographs, vol. 13. VCH, Weinheim
19. Scheper T, Reardon K (1992) In: Sensors – a comprehensive survey, vol. 3(2). VCH, Weinheim, p 1023
20. Scheper T (1995) Nachr Chem Tech Lab 43:328
21. Fillipini C, Sonnleitner B, Fiechter A, Bradley J, Schmid R (1991) J Biotechnol 18:153
22. Oehme F (1992) CLB Chem Lab Biotech 43:309
23. Lüdi H, Garn MB, Bataillard P, Widmer HM (1990) J Biotechnol 14:71
24. Manz A, Graber N, Widmer HM (1990) Sensors and Actuators B 1:244
25. Ocvirk G, Verpoorte E, Manz A, Grasserbauer M (1995) Analytical methods instrumentation, vol 2, no 2
26. Burggraft N, Manz A, Verpoorte E, Effenhauser S, Widmer H M, de Rooij NF (1994) Sensors and Actuators B 20:103
27. Fettinger JC, Manz A, Lüdi H, Widmer HM (1993) Sensors and Actuators B 17:19
28. Van der Linden WE (1986) Anal Chim Acta 179:91
29. Ruzicka J, Hansen EH (1984) Anal Chim Acta 161:1
30. Garn M, Gisin M, Thommen C (1989) Biotechn Bioengineering 34:423
31. Verpoorte EMJ, van der Schoot BH, Jeanneret S, Manz A, Widmer HM, de Rooij NF (1994) J Micromech Microeng 4:246
32. Becker EW, Ehrfeld W, Hagmann P, Maner A, Münchmeyer D (1986) Microelectron Eng 4:35
33. Manz A, Verpoorte E, Raymond DE, Effenhauser CS, Burggraf N, Widmer HM (1995) In: van den Berg A, Bergveld P (eds) Micro total analysis systems. Kluwer Academic, p 5
34. Manz A, Graber N, Widmer HM (1990) Sensors and Actuators B 1:244
35. Drobe H (1992) GIT Fachz Lab 10/92:1034
36. Manz A, Harrison DJ, Verpoorte E, Widmer HM (1993) Advances Chromatography 33:1
37. Snyder LR, Kirkland JJ (1979) Introduction to modern liquid chromatography. Wiley, New York
38. Jandik P, Bonn G (1993) Capillary electrophoresis of small molecules and of ions. VCH, New York
39. Effenhauser CS, Paulus A, Manz A, Widmer HM (1994) Anal Chem 66(18):2949
40. Raymond DE, Manz A, Widmer HM (1994) Anal Chem 66(18):2858

6 Metallurgical Process Technology and Chemical Process Analytics

6.1
Introduction

6.1.1
Economic Importance of Iron (Steel) and its Metallurgy

This chapter should be preceded by a poet's epigram, being characteristic and still valid today (Friedrich von Logau, 1654):

Das Eisen dünkt mich	Iron, I do think
Ist weit mehr als Gold zu preisen;	Is to be praised far more than gold;
Ohn' Eisen kommt nicht Gold	Without iron comes not gold,
Gold bleibt auch nicht ohn' Eisen.	Nor does gold stay without iron.

Among the industrially produced metals, due to the per head consumption in the world and the great variety of steel grades [1] the metal iron occupies a special position [2]. Therefore it is advisable, in an observation of the relationships between modern, efficient process technology and process analytics closely linked with it, to emphasize steel as being an economically and scientifically particularly interesting representative of metallic materials and, with the help of this example, to make clear the variety of process analytical tasks.

The economic importance is explained by the fact that iron and steel are still a dominating factor in our times. Of all the metals which are obtained industrially, iron is, with a share of 95%, the most widely used. And yet world-wide requirements are steadily increasing. The "iron age" is therefore not yet over despite numerous developments and efforts made in other material fields such as those of ceramic materials and plastics which are leading to a variety of substitution tendencies. Due to application-oriented further developments and the creation of new grades of steel, steel is still a modern material. Its great adapt-

---

1 Iron materials with less than 1.7% m/m carbon are called "steel". In Middle High German it is "stal", "stahel", Old High German "stahal", German "Stahl". All these words mean hard and strong.
2 This is deliberately simplified; there are, as it is well known, a large number of high alloy steels ("super alloys") which contain only a small amount of iron.

ability and wide application purposes become clear as a result of, among other things, the following figures: there are about 2000 German standard and regular steels, and almost 1000 steel grades according to foreign and international standards which are supplemented by about 7000 "special works qualities". Accordingly, steel is the most versatile material which exists, and can be considered a material that can meet the technological challenges of today and those of the future. Also, its future importance cannot be reduced by the fact that the steel industry in many countries is in a structural crisis and that classical producer countries have had to accept considerable losses in crude steel production in recent years.

In fact, alternative materials such as aluminum and plastics are used in many areas of life. However, their applications are limited above all because of their mechanical-technological properties. Each automobile contains, for example, about 80 kg of plastics but no significant new applications of plastics for automobiles can be expected at this time. Similar facts apply to the use of ceramic materials in motor construction. In this case, development is fully underway, but a quantitatively significant substitution of steel materials is still not foreseeable at the present time.

There are a number of arguments in favour of steel with relation to production and process analytics which will be dealt with in more detail for the reasons stated in this introduction. The "uniqueness" of steel is first of all to be verified with some examples.

Petroleum and natural gas are today extracted from depths of up to 8000 m. These boring and conveying depths expose the whole steel structure of the boring platform, and also the steel tubes used, to increasingly higher stresses which result from tension, external and internal pressures, bending and seawater corrosion. Aggressive media such as acidic gases, carbon dioxide as well as chlorides also require extreme corrosion resistance. The opening up of energy sources in the off-shore area would not have been possible without the development of new, high strength, weldable fine-grained steels. The advantages of these steel grades are also used in pressure vessels and crane construction.

The positive relations of the material steel with the topic "environment" are mixed. First, iron is a natural element which will still be available for a very long time in secured deposits. Second, iron degrades to iron oxide due to corrosion, which in fact is economically not very meaningful but is ecologically safe. And the most important point – steel is the only material which can be _reused_ without problems and for which a closed, proven and functioning recycling system exists. Besides the technical efficiency of a material its later disposal will gain increasingly greater importance in all countries of the world. The materials of the future must be fully recyclable. Recyclability thereby refers to the recovery of the material from goods and plants which are no longer functioning as well as its subsequent _reuse_ and the _reuse_ of production waste.

The growth of world steel production today is no longer the domain of the classic producing countries, but increasingly of countries which in the past had only a small amount of steel production (e.g. China, India, Australia, Spain, Rumania, South Africa, Hungary) or no steel production of their own worth mentioning (e.g. Brazil, Mexico, Argentina, Korea, Bulgaria). From the latter group, Brazil, which has particularly promoted its steel industry with an aggressive, growth-oriented economic policy, is picked out as an example. Thus by international comparison, Brazil is, with an annual crude steel production of approximately 25 million tonnes, in seventh place in the list of industrialized countries. This last-mentioned development naturally has an effect on the *sociological structure* of the countries and influences their *standard of life*. Because of this relationship one frequently sees the per head steel consumption of the population of a country as the measure of the average standard of life. As a consequence of the development outlined, a percentage displacement in world production occurred. While about 35 years ago the USA, Germany, the United Kingdom and France together held a 65% share of world production, this share is today only about 20%.

6.1.2
Presence of Iron in Nature

In accordance with its frequency in the earth's crust, iron is a constituent of numerous minerals, chiefly occurring bound as oxide, sulfide, carbonate and silicate. Called as iron ores[3] these deposits are the starting point from which the industrial production of iron takes place[4]. They are not pure minerals, but are mixtures of ferrous compounds and non-ferrous accompanying minerals ("gangue"). The most important ores contain the iron in an oxidic form; today carbonaceous ores are no longer of importance (e.g. Styrian Ore Mountain). In 1990 a total of about 970 million tons of iron ores were mined in 50 countries around the world.

As a rule, the gangue consists mainly of compounds containing lime, silicic acid, alumina and phosphate. As the phosphorus is fully transferred to the pig iron during the metallurgical treatment of the ores in the blast furnace, the process for steel production must be composed in such a way that it permits removal of the unwanted phosphorus to a large extent (see below). Thus, the first process analytical task for the metallurgy of iron becomes apparent.

The oxides Fe_2O_3 and Fe_3O_4 involved in the genesis of oxidic iron ores tend, like the third iron oxide FeO, to the formation of non-stoichiometrically constituted phases; that means that in their crystal lattices all the lattice places are not always completely occupied with Fe ions. Iron(III) oxide occurs, depending on the way it has been formed, in a water-containing form ($FeO(OH)$) or a form free of water (Fe_2O_3). On heating up to 200 °C the brown $FeO(OH)$ containing water

---

3 "Ore" probably comes from the Babylonian or Sumerian word urud = copper. In Old High German it was aruzzi, arizzi, aruz and in Middle High German erze, arze.

4 Definition of the time-dependent term "iron ore" – a material containing iron which is currently used and will in the foreseeable future be used taking into account the local conditions and the current and future foreseeable costs of and the market prices for the production of iron and steel.

passes into the reddish-brown Fe_2O_3 (corundum lattice) which in nature is found as *haematite* and forms the main constituent of kidney ore. Ferromagnetic Fe_3O_4 which appears in nature as *magnetite* is an iron(II/III) oxide. It is formed, for example, also during the oxidation of iron ("scale", "cinder"). However, iron(III) oxide does not only form double oxides, the "ferrites", with iron(II) oxide but also with numerous other metal oxides. The spinel-type compounds with bivalent oxides, such as ZnO, MgO, NiO are of great industrial interest because of their electrical and magnetic properties. Magnetite can also be considered – although with an inverse structure – a spinel: $Fe^{3+}[Fe^{2+}Fe^{3+}O_4]$, i.e. one half of the Fe^{3+} ions in the spinel lattice occupy the tetraeder positions and the other half together with the Fe^{2+} ions occupy the octaeder positions. The electrical conductivity is based on the electron transfer between the ions located in equivalent lattice places.

6.1.3
History

Iron[5] appears as a commodity metal (iron age) in various parts of the world at different times [1]. While the iron age starts in Egypt and India in the 6th century BC, in China and Japan iron was first produced in the 2nd century BC. In Europe the production of iron dates back to the first millennium BC. It is worth mentioning that not all cultures have become acquainted with iron metallurgy in their development. Weapons and agricultural implements are always the first iron products. In prehistoric times easily reducible iron ore was mixed with an excess of charcoal and reduced in a natural air stream (direct process for the production of wrought iron). During this process a solid to doughy product (bloom) was obtained which was freed of slag by repeated forging.

The growing demand for iron (steel) led to the development of special furnaces in which the iron ores could be reduced in larger quantities and in which finally liquid pig iron was obtained, and later to furnaces in which the pig iron obtained could be freed of its unwanted accompanying elements by means of oxidizing gases (refining). When charcoal was no longer able to meet the growing demand of the iron and steelworks at the beginning of the 18th century, hard coal coke was used as a reducing agent. However, there is, interestingly, in some countries for geo-economic reasons, a "revival" of the charcoal technique, such as in Brazil where today one third of the pig iron is produced using charcoal.

The progress of civilization and the associated steel consumption led to the development of new steel production processes. First of all, the *air refining processes* should be mentioned in which air is blown through the pig iron (H. Bessemer, 1855; S.G. Thomas, 1878). Besides that, the *Siemens-Martin process* was

---

5 The word "iron" comes from the Celto-Illyrian word "isarno". The Celtic original produces the Gothic "eisarn", Old High German "isarn", Middle High German "isen", Anglo-Saxon "iron". Names such as Isambart, Isolde, Isegrim, Iserlohn (wood in the iron ore region), Isenburg etc. can be traced back to this word.

developed in which the oxidizing process is carried out in a hearth type furnace by means of flame gases (P. Martin, 1864). With the turn of the century, steel production in *electric furnaces* (F. A. Kjellin; P. Heroult) begins, while in more recent times various *basic oxygen steel processes* have been developed in which pure oxygen is blown onto or through the liquid pig iron. Thus, a technology is arrived at which will have to be looked at in more detail from the point of view of process analytics [2] and which, as a result of its high efficiency, enables the economic developments of our times. The clarification of the metallurgical processes taking place during the production of iron and steel was only possible in recent times due to the development and application of physical chemistry.

6.1.4
The Metallurgical Products Pig Iron and Steel

In order to understand better the following remarks on chemical process analytics the material steel and its main base product pig iron are first to be described. The term *steel* is applied to a group of deformable iron materials which extends from unalloyed, industrially pure iron up to high alloy iron with carbon contents of up to a maximum of 1.7% m/m [3]. On the other hand, *pig iron* is an intermediate product in steel production high in carbon (3.5–4.5% m/m C). *Cast iron* is chemically related to it. The carbon chemically bound by the iron (iron carbide or "cementite", Fe_3C) and the carbon partly occurring freely in the iron (graphite) influence the melting and transformation temperatures, the microstructural composition and thus the most important properties (strength, dilatation, hardness, formability, weldability). If all the carbon is bound as carbide one speaks of white, otherwise of grey pig iron, since in the latter case the majority of the carbon is present as graphite and this gives the surface of fracture a grey appearance. The microstructure of pig iron and steel shown in Figs. 6.1 and 6.2 makes clear the difference in structure of these substances and enables the understanding of the different properties.

Pure iron is not a useable material for most applications so other elements must be added to it in precisely defined proportions ("alloying") in order to achieve certain technological properties which can vary within wide limits. For alloying mainly mixed metals (ferroalloys) are used which usually contain the required alloy component as the predominant constituent and have to be analytically evaluated before use. The effect of the alloy metals varies widely, whereat the simultaneous use of several metals does not always produce the sum of the individual effects. For the final material properties, moreover, the processing and heat treatment are of decisive importance. In accordance with the variety of demands made on steel as a material for almost innumerable requirements, the amounts of the alloy elements can differ very widely. A distinction is made between unalloyed, low alloy and high alloy steels.

In the case of the unalloyed carbon steels which are used as construction and tool steels, the proportions of the "alloy components" are in each case below 1% m/m, e.g. for Mn<0.8% m/m, Si<0.5% m/m, Cu<0.25% m/m, Ti<0.1% m/m, Al<0.1% m/m (these data correspond to the German standard DIN 17006). The

Fig. 6.1. Microstructure of pig iron (magnification 100✕)

Fig. 6.2. Microstructure of carbon steel (magnification 200✕)

properties of these steels are, on the whole, determined by the carbon content (0.1–1.2% m/m). With increasing C content, for example, the tensile strength increases (strength against tensile stress)[6] while the ductility (plastic formability)[7] and the forming property, e.g. by rolling, decrease.

Low alloy steels contain up to 5% m/m alloying components. Above this limit the steels are high alloy steels.

When alloyed steel is referred to in everyday speech, one first thinks of the well-known corrosion-resistant chromium-nickel steels which have made inroads into daily life over a wide area. However, this field is considerably more multi-faceted. In low alloy steels one finds either alone or together mainly Cr, Cu, Mo, V, Ni, Mn and Si. The high alloy steels contain Cr and/or Ni, and in addition often Mo, V, Nb, Ta, Ti and W. In order to make it easier to understand the numerous analytical tasks which are posed here and which involve a reliable determination of numerous elements in high amounts as well as in the trace range, some groups of materials are briefly described below.

By means of the alloy metals, *high strength steels* (up to 18% m/m Cr, up to 18% m/m Ni, up to 5% m/m Mo plus, if necessary, small amounts of W, Cu, Nb, Ta, Al, Ti, Mn, Co, B, Zr) are successfully produced, with an application range from –200 °C up to +700 °C and above. *Heat resistant steels* are steels which at temperatures above about 550 °C have an increased resistance to attack by gases. The "non-scaling property" (resistance against oxidizing gases) can extend up to 1200 °C. The contents of alloy constituents: 6.5–29% m/m Cr, 0–35% m/m Ni, up to 2.3% m/m Si, up to 1.5% m/m Al. The *corrosion-resistant* steels are mostly chromium and chromium-nickel steels with at least 12% m/m Cr which, in order to achieve certain material properties, can contain further alloy additions. There are corrosion-resistant steels with up to 26% m/m Cr, up to 26% m/m Ni and up to 5% m/m Mo, as well as up to 2.5% m/m Si, up to 20% m/m Mn and up to 2.5% m/m Cu. In the case of steels particularly resistant to sulphuric acid, the Ni content can even amount to 42% m/m with the simultaneous addition of Mo and Cu. The individual steels of this group in each case require special heat treatment (achievement of a particular microstructure) in order to obtain the required service properties.

As the last group the *tool steels* are to be mentioned. By that, one understands steels which are used for the production of tools in the widest sense (for hot and cold forming, dividing and crushing of the most varied materials; parts for measuring instruments). Besides unalloyed tool steels there are alloyed steel grades and the "high-speed steels" which enable particularly high speeds during "cutting" machining. The alloy elements are combined in very different ways in the

---

6 *Tensile strength*: this refers to the maximum force occurring during the tearing of a sample, based on the initial cross-section. The tensile strength is found by dividing the tensile force by the sample cross-section. The tensile strength of a material is determined by a tensile test.

7 *Ductility*: the ability of the material to be plastically deformed. The microstructure tries to postpone destruction by a high degree of deformation (fracture on working in contrast to cleavage fracture of brittle materials). A measure of ductility in the case of static stress is the percentage elongation after fracture, in the case of impact or stroke the impact strength.

alloyed tool steels. Their contents extend up to 12% m/m Cr, up to 18% m/m W, up to 8% m/m Mo, up to 4% m/m V and up to 3% m/m Co with manganese and silicon contents of up to 3% m/m and 2% m/m, respectively. The C contents are >0.5% and reach 2.3%. In this material group, heat treatment is naturally of great importance (high-temperature strength, wear resistance, hardness). The high-speed steels are on the borderline with "hard metal alloys" (cutting metals); the latter consist essentially of 30–50% m/m Co, 25–30% m/m Cr and 6–20% m/m W.

In conclusion it must be pointed out that this short presentation cannot be complete. It was intended that some main material groups were to be picked out. In particular, however, it was necessary to convey a first impression of demands made on the analyst (process analyst) which are directly derived from the variety of the (steel) materials. In order to put some order into the bewildering variety of steel grades, steel standards, which specify the chemical composition and the mechanical properties of the individual steel grades, were developed by co-operation between the steel producers, the steel trade and the steel consumers, with the involvement of scientific institutes [4]. These standards – there are those which have national, European and world-wide validity – are binding on the steel manufacturer by inclusion in supply contracts. They are meant to guarantee for the consumer the constant quality and the required properties (see Chap. 10).

6.2
Process Analytics of the Blast Furnace Process

6.2.1
Metallurgy and its Analytical Requisites

The blast furnace process is currently, and will in the foreseeable future be, by far the most economical way to convert iron ores into pig iron and thus subsequently into crude steel. About 95% of the ores are reduced to liquid pig iron in blast furnaces which are typical countercurrent reactors. The development in blast furnace technology in recent years was characterized by the construction of large volume blast furnaces with a daily production of up to 10,000 tonnes. These are plants which can be fed and controlled largely automatically.

The metallurgy of iron and steel production must, because of the essentially oxidic composition of the iron ores and of the metallurgical slags, be regarded as *high temperature chemistry of oxides*. Besides the main constituent, iron, and the already mentioned fluxes for making slag, the iron ores contain low amounts of a number of other elements which automatically pass into the pig iron produced in the blast furnace process (accompanying elements to iron). Their contents can, depending on the ore basis and the process conditions, vary within wide limits. One can distinguish between desired and undesirable main accompanying elements ($5-10^{-1}$% m/m), secondary accompanying elements (order of magnitude 10^{-2}% m/m) and trace elements (order of magnitude 10^{-3}% m/m).

Manganese, silicon and phosphorus can be stated as being main accompanying elements. They are oxidized during the subsequent steel production and supply during the process, together with the carbon contained in the pig iron, the thermal energy needed for the (basic oxygen) process. The main accompanying element carbon originates from the blast furnace coke, fuel oil or natural gas used or from the coal injected into the blast furnace. It is from the fuels that the greater part of the undesirable accompanying element to iron, sulphur, comes (S contents: coke 0.7–1% m/m ; fuel oil 0.6–1% m/m).

Among the secondary accompanying elements, which the blast furnace operator can influence within certain limits by appropriate selection of raw materials, are Cu, Cr, Sn, Ni, V, As and Ti. Their amounts are in the order of 10^{-2}% m/m to 10^{-3}% m/m. In the pig iron required for steel production these elements are mostly undesirable as they unfavorably influence the properties of many steels; thus, for example, higher Cu contents affects the rollability and cause – similar to sulfur – red shortness[8] of the iron. Steel plates which are subjected to particularly strong forming (deep drawing[9]) have to contain not more than 0.03% m/m Cu. This circumstance requires appropriatcly low contents in the pig iron used as the base material. Tin becomes concentrated under layers of scale and causes solder brittleness (cracks)[10]. Arsenic increases the temper brittleness[11], reduces the ductility and worsens the weldability.

Elements which occur in pig iron in the order of 10^{-3}% m/m, are referred to as traces[12] . These include, for example, Al, Co, Pb and Zn. Even by these low amounts the properties of pig iron and cast iron are strongly influenced. Use is made of this fact in the foundry industry where additives of Bi, B, Mg and Sb of the order of 10^{-2} up to 10^{-3}% m/m are of technical importance. These few notes provide evidence of the importance which is attached to chemical analytics in the selection of raw materials and process control.

Blast furnace process is a continuous shaft furnace process in which the "burden" (ore mixture with additions) and the coke required for reduction and as a thermal energy source "flows" towards a hot gas stream and in which, thus, an intensive heat and matter exchange between gaseous and solid phases is ensured (*countercurrent principle*).

Now the process sequence should be followed from the furnace top to the hearth regarding the running of the solids (Fig. 6.3) in order to be able to recognize the importance of the process analytical tasks to be dealt with below.

In the upper part of the furnace (shaft) the cold "charge" is heated up by the counter-flowing gas to 400–500 °C and thus the moisture adhering to the feed-

---

8 *Red shortness*=tendency to bursting of the red hot steel (800–1000 °C) during forming.
9 *Deep drawing*=forming of plane metal sheets to hollow bodies by means of stretch forming and pressure forging, in certain circumstances in several work cycles. The deep drawing property depends essentially on the purity of the steel (low C content, little contamination) and the surface structure of the metal sheet.
10 *Solder brittleness*=loosening of the microstructure by penetration of easily melting, liquid metals (solder) into the grain boundaries which leads to intercrystalline cracks.
11 *Temper brittleness*=embrittlement of certain steels after heat treatment (tempering) and subsequent slow cooling.
12 The term "*traces*" is not always clearly defined. In chemistry, considerably lower concentrations are usually called traces (range 10^{-4} to 10^{-12}% m/m).

stock is evaporated. In addition, the water obtained from the decomposition of hydrates escapes. With a further temperature rise the carbonates charged with the burden are decomposed, whereby carbon dioxide is formed.

In the following temperature zone up to 1000 °C (lower part of the shaft) the reduction of the ore by the carbon monoxide generated from the coke (indirect reduction) takes place. However, the reduction does not proceed in one reaction up to metallic iron, but is accomplished in several stages in parallel and one after the other. That is understandable if one bears in mind that the various oxides at the same temperature have different decomposition pressures (oxygen partial pressures) and thus different stabilities. As the final oxide stage "wuestite", $FeO_{1.06}$, is formed, which represents a solid solution of FeO and Fe_2O_3.

The production of the CO necessary for the reaction starts in the lower part of the furnace (hearth). During oxidation of the coke with air blown into the furnace at a temperature of approximately 1200–1300 °C ("blast") carbon dioxide is initially formed which is then converted endothermically into CO (Boudouard reaction):

$$CO_2 + C \underset{>1000\,°C}{\rightleftharpoons} 2\,CO$$

The state of equilibrium of the Boudouard reaction depends on the temperature and lies, at high temperatures in the hearth, completely on the CO side.

In the zone of indirect reduction (in the shaft <1000 °C) the CO_2 formed during the reduction of the ores is no longer able to react with the coke. By contrast, the CO_2 formed at higher temperatures (>1000 °C) in the bosh parallels and in the bosh of the blast furnace immediately reacts again, according to the Boudouard reaction, with the coke producing CO which in the further course of the process is again able to reduce iron oxide:

$$C \;+\; CO_2 \;\Rightarrow 2\,CO$$
$$FeO \,+\, CO \Rightarrow Fe \,+\, CO_2$$
$$\overline{}$$
$$FeO \,+\, C \;\;\Rightarrow Fe \,+\, CO$$

Since, looking at the total process, solid carbon is converted so that this region is called the zone of "direct reduction" [13]. A "genuine" direct reduction due to the direct effect of carbon only occurs to a small extent and thus does not have for the production of iron the same importance as for the production of other commodity metals (e.g. copper, zinc).

Moreover, of special importance for the blast furnace process is the reversal of the Boudouard reaction, i.e. the decomposition at the low temperatures (<1000 °C) of unstable carbon monoxide, catalytically accelerated by the initially produced, finely dispersed metallic iron. The carbon be formed in this way can already react in the solid phase with the iron (diffusion) and in the course of the

---

13 *Robert Bunsen* (1838), the co-founder of emission spectral analysis, had already succeeded in explaining the chemical processes in the blast furnace.

process is solved to a considerable extent from the iron ("carburization"=formation of Fe_3C) and lowers its melting point to about 1200 °C.

These processes enable the pig iron to be discharged ("tapped") in low-viscosity form from the furnace (hearth) after comparatively little overheating.

The water vapor always contained in the "blast" (hot air) reacts with the incandescent coke forming hydrogen ("*water gas reaction*"). The latter participates in the further course of the process in the reduction of the iron oxides. More recently greater importance has been attached to the hydrogen as a reducing agent in the blast furnace, as part of the coke is in many cases replaced by the addition of water vapour to the blast and by the injection of coal and/or petroleum.

In the course of the processes described, from the gangue of the ores and the ash constituents of the coke, *blast furnace slag* is formed (Table 6.1), consisting mainly of calcium silicates and aluminates. In order to be able to remove these from the process in a form with the lowest possible viscosity the "acidic" (SiO_2, Al_2O_3) and "basic" (CaO, MgO) components must be in a certain ratio to each other (e.g. CaO/SiO_2=1.2).

At the end of the process the *pig iron* is obtained which, when charging ores low in phosphorus, has roughly the composition given in Table 6.2 and is used as the base material for subsequent steel production.

Table 6.1.
Chemical composition of blast furnace slag (example)

Component	Amount % m/m
Calcium oxide	41
Magnesium oxide	7
Silicon dioxide	36
Aluminum oxide	11
Titanium(IV) oxide	0.4
Iron	0.15
Manganese	0.15
Sodium oxide	0.3
Potassium oxide	0.7
Sulfur	1.3
Phosphorus	0.01
Nitrogen	0.025

Table 6.2.
Chemical composition of hot metal (pig iron) (example)

Element	Amount % m/m
C	4.5
Si	0.5
Mn	0.35
P	0.09
S	0.008
Cu	0.006
Cr	0.02
Ni	0,008
Ti	0.02
V	0.008
N	0.005

6.2.2
Process Technology and Process Analytics

The economical production of pig iron depends not only on the charge material which is optimum with regard to its chemical composition and its physical properties but also on process-oriented metrology and process analytics. The process and the continuous monitoring of the production of iron ore sinter, a preferred base material particularly suitable for use in the blast furnace, has already been dealt with in Sect. 4.3.4. The process analytical tasks during sinter production are summarized in Table 6.3.

The sinter is charged together with lump ores, other Fe-containing materials and the blast furnace coke necessary for reduction into the blast furnace, a shaft furnace, already described (see Fig. 6.3). Depending on the furnace capacity and the hearth height the pig iron is "tapped" (temperature about 1400 °C) every 2–4 h, when the quantities obtained during each run-off can amount to several hundred tons. Since on large blast furnaces several tap holes can be operated one after the other, one can talk in that case of a quasi continuous tapping of the pig iron, which, as the primary material for the subsequent steel production, has to be analysed without any time lag.

Figure 6.4 gives a general idea of the present-day state of metrology and process analytics considered necessary, which is the prerequisite for the development of models for process optimization and control of the thermal economy and thus for high performances. It becomes apparent that in the case of this first metallurgical stage a greater role is attributed to chemical process analytics than in the case of raw material processing (cf. Table 6.3). Therefore some interesting aspects and more recent analytical developments are to be looked at more closely below.

Important control variables for the optimization of the thermal economy are obtained from the analysis of the top gas, which is used as a fuel gas (composi-

Table 6.3. Process analytics for sinter production

	Component	Analytical method	Aim of investigation
Fuel			
(Coke breeze)	Water	Neutron scattering	Constancy of ballast material proportion and real fuel content
	Ash	γ-ray backscattering	
Raw material			
(Sinter blend)	Si, Ca, Fe	X-ray fluorescence	Homogeneity and quality of mixing
	Water	Neutron scattering	
		IR Spectrometry	
Product			
(Sinter)	Si, Ca, Al, Ag, Ti	X-ray fluorescence	Regulation of basicity
	Fe^{2+}	Magnetic permeability	Reducibility

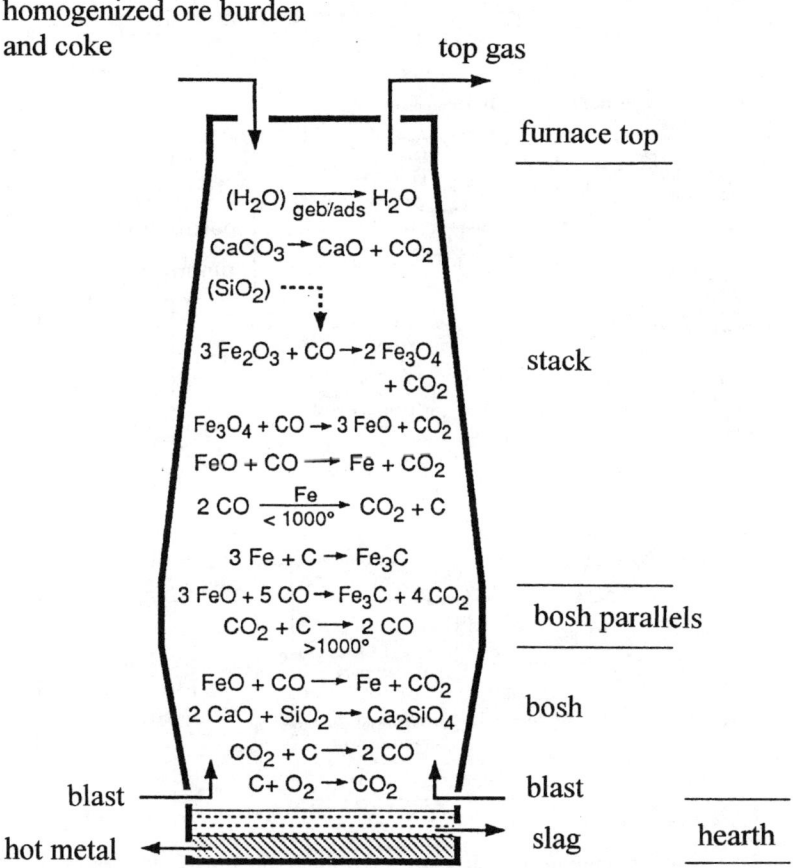

Fig. 6.3. Chemistry of the blast furnace process

tion approx. 22 vol. % CO_2, 21 vol. % CO, 2 vol. % H_2, remainder N_2; gross calorific value: 2850 kJ/m³ (STP)). In this case, for the continuous determination of the CO_2 and CO amounts, proven infrared spectrometric methods have been in use for a long time, while the H_2 amount is obtained by means of a thermal conductivity measurement, the measuring procedure including the calibration and adjustment procedures being computer-controlled (see Sects. 2.1 and 2.2). Attempts have repeatedly been made to make use of mass spectrometry for this part of process analytics.

The most important task naturally consists of the characterization of the main product obtained, "pig iron". That has so far been done by sampling during tapping and from the "transport ladles" and their subsequent spark emission or X-ray fluorescence spectral analysis (see Sects. 4.2.1 and 6.3.2). The samples in that case as a rule are passed through a pneumatic tube conveying system into the analytical laboratory. Pig iron samples taken from the molten metal bath

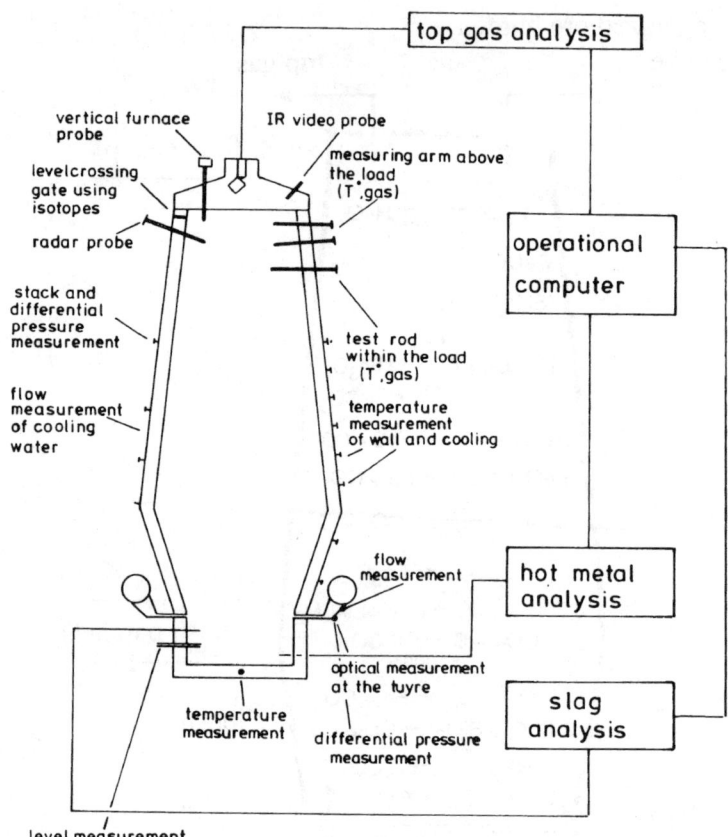

Fig. 6.4. Scheme of the equipment for measuring technique and process analytics at a blast furnace

show, by comparison with steel samples, a coarser, more heterogeneous microstructure (see Sects 6.1.4 and Fig. 6.1). Graphite precipitation and local element concentrations (segregation) can completely exclude a spark emission spectrometric analysis. Therefore in that case either by a special sample shape, which permits rapid cooling of the sample, or by the remelting of the primary sample with subsequent quenching, a "white" solidified analysis sample must be obtained in which the carbon is present homogeneously distributed as iron carbide. In the case of the latter method the sample is melted under protective gas by means of high-frequency heating and by centrifuging ejected into a copper ingot mould [5]. The high temperature gradient guarantees homogeneous solidification. In the case of remelting of graphitic pig iron with a carbon content of more than 4.3% m/m, with a silicon content of 1–2% m/m and a manganese content of less than 0.75% m/m, samples are obtained which can be analyzed without any problems on vacuum emission or X-ray fluorescence spectrometers. In the first case, the determination of carbon is also possible.

Table 6.4.
Precision (repeatibility) of the XRF spectrometric analysis of hot metal and blast furnace slag

Hot metal (pig iron)		
Element	Range % m/m	Standard deviation s % m/m
Si	<0.5	0.003
Mn	<0.2	0.004
	0.2–2.0	0.006
P	<0.1	0.002
	0.1–1.0	0.006
S	<0.1	0.002
	0.1–0.25	0.006
Cu	<0.1	0.002
	0.1–1.0	0.006
Cr	<0.1	0.002
	0.1–1.0	0.008
Ni	<0.1	0.002
	0.1–1.0	0.008
Ti	<0.1	0.002
	0.1–0.5	0.003
V	<0.1	0.003

Blast furnace slag		
Component	Range % m/m	Standard deviation s % m/m
SiO_2	0–50	0.2
Mn	0–1	0.05
P	0–0.2	0.005
S	0–2.5	0.01
Al_2O_3	0–17	0.05
TiO_2	0–1.5	0.005
Fe	0–2.5	0.008
CaO	25–50	0.2
MgO	0.5–18	0.05
Na_2O	0–1.0	0.02
K_2O	0–2.0	0.02

If the determination of the carbon can be renounced, the use of X-ray fluorescence spectrometry can be considered as it makes lower demands on the quality of the samples. (In the case of carbon-saturated pig iron, with the help of a model calculation using the silicon content and the temperature of the liquid pig iron, the carbon content can be determined with the same reliability as by spark emission spectrometry.) As for process control, the chemical composition of the blast furnace slag, which is usually determined by XRF, is also of significance for the control of blast furnace operation; in this case the analysis of the main product, pig iron, *and* of the reaction product (by-product), blast furnace slag (see later), can be carried out with a single method.

This circumstance makes it easy to start with automation of process control [6]. The samples arriving via a pneumatic tube system are led on the basis of an identification signal to two different preparation lines. Such a system consists of an automatic belt grinder for the pig iron sample, an automatic slag preparation machine and a multi-channel X-ray fluorescence spectrometer with sample changer. The data are automatically transferred to the blast furnace plant after a plausibility test by the laboratory computer. In order to control the complete analytical system and for the recalibration of the analyzer a terminal with all operating functions is available to the personnel working far off in the central analytical department. The maintenance schedule envisages, for example, a 24-h cycle for the cleaning of the machines and for the replacement of the abrasives.

Regarding the technique of sampling and spectrometric analytics reference should be made to the notes in Sects. 6.3.2.1 and 6.3.2.2. As analytical characteristic data for this field of process analytics, the standard deviations (2 s) for the XF spectrometric analysis of hot metal and blast furnace slag are listed in Table 6.4.

Even though sample preparation and analysis were largely automated due to the demands of the metallurgists, because of the increased requirements, systematic, quality-oriented production was directed towards a "delay-free" determination of the most important elements useable as control variables. The element amounts meeting the comprehensive quality requirements could, if on-site methods were used, still be determined in the well-known ways, e.g. by X-ray fluorescence spectrometry in a central laboratory.

It is as a process analytical contribution to this requested improvement in blast furnace process control that the work on "direct determination" of the elements Si, Mn and S in the liquid metal should be assessed, which could open up the possibility of in-line or on-line analysis. The following spectrometric and electrometric methods are tested.

1. *Aerosol generation by spark excitation and ICP spectrometry of the aerosol*
   In this procedure, developed in Japan [7], by means of an electric discharge (Fig. 6.5) particles are generated which are transmitted in an argon flow into the plasma torch of an ICP spectrometer located 40 m (!) away. Operating tests in a pilot plant which revealed considerable investment and maintenance expenditure were carried out.
2. *Direct spark emission spectrometry*
   In another approach [8] material is evaporated and excited by means of an electric spark and the light emitted is transferred to a spectrometer with the help of a light guide. Under laboratory conditions concentration-dependent measuring signals were received in the range 0.1–0.8% m/m Si. The question of the influencing of the process, e.g. by existing slag phases, has not yet been conclusively clarified.
3. *Direct laser emission spectrometry*
   Moreover, there are investigations [9] to utilize spectrometrically the direct excitation of the liquid metal by means of a laser. The first results show that laser emission spectrometry seems to be suitable for the analysis of hot metal. It could be shown that for the elements C, Si, Mn, P and S values resulted which were in satisfactory agreement with the data obtained on samples taken.

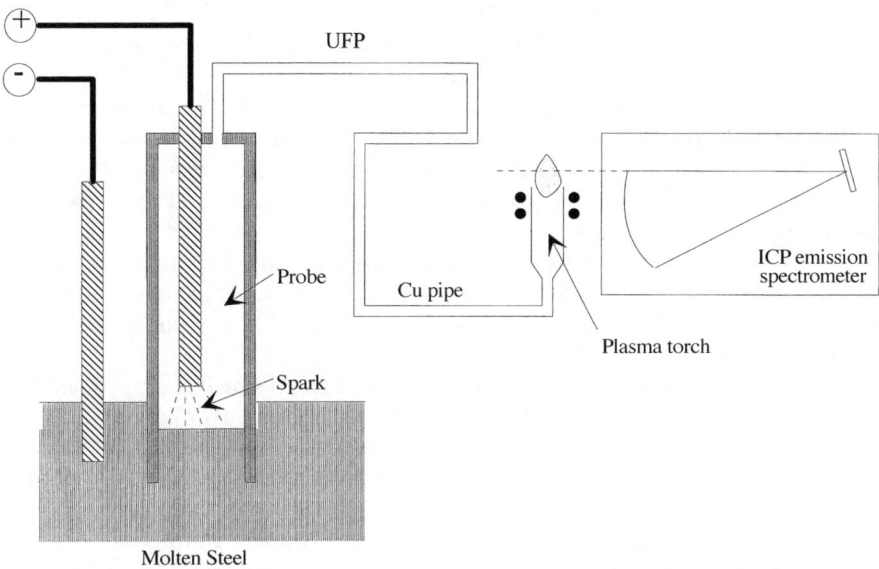

Fig. 6.5. Scheme of an aerosol generator for the in-line analysis of hot metal

4. *EMF measurement for Si determination*

A development in the sense of sensor technology [10] is the electrochemical determination of the Si content in molten pig iron. This probe corresponds to a galvanic silicon concentration cell of the type:

Ni_{sat}-Si (s)//silicate electrolyte(s) // Fe-C-Si (s),

and is suitable for measurements in C-saturated Fe-C-Si melts at 1400 °C and with Si contents of 0.05–1.8% m/m.

In recognition of the fact that the determination of the Si content in the pig iron nowadays can no longer serve alone as a process indicator, research work has been taken up for the development of further electrochemical sensors, e.g. for sulphur. Electrochemical investigations on CaF_2-CaS solid ion conductors for the measurement of sulfur activities have led to first approaches [11].

If one considers the production of pig iron a preliminary stage of modern steel technology, a glance at examples requiring adjustment of amounts of some lowest accompanying and trace elements is interesting. The mean amounts of trace elements have decreased substantially over the past 30 years:

	1995 % m/m	1965 % m/m
Cu	0.005	0.04
As	0.005	0.025
Sn	0.005	0.01

In the case of non-metallic accompanying elements, after a secondary metallurgical treatment, values of sulfur <0.007% m/m, of phosphorus <0.008% m/m and of nitrogen of about 0.001% m/m can be achieved.

Thus, these proportions are in an order of magnitude which already requires particular analytical measures and methods and which only a few years ago would have caused problems with regard to reliable determination. Particular attention has to be given to the calibration of the analyzers and the creation of suitable reference materials.

From the gangue of the ores, the ash constituents of the blast furnace coke and the auxiliaries, the blast furnace slag is formed (for an example of the composition of a blast furnace slag when producing pig iron with low phosphorus content, see Table 6.1).

It is usually poured into "slag beds" in which it cools slowly. Lumpy blast furnace slag is an excellent road construction material due to its high compression strength. If the liquid slag is allowed to flow into water tanks it solidifies to small grains. The "slag sand" thus obtained is used as a constituent of slag cements and for the production of slag bricks. "Foamed slag" and "slag wool" are heat and sound insulating materials. The former is produced by the foaming of slags high in silicic acid with water to produce "pumice stones". Slag wool is produced by atomizing liquid slag with the help of a steam jet. Blast furnace slags, ground fine as dust and with a high lime content, are called "slag lime" (fertilizer).

In all these cases mentioned, chemical analytics plays an important process technological (see above) and commercially relevant role.

6.3
Process Analytics of Steel Production

6.3.1
Metallurgy and its Analytical Presuppositions

The hot metal produced in the blast furnace is the main charge material for the second metallurgical process stage, steel production. At present there are two process approaches of primary interest, the basic oxygen steelmaking process (BOS) in its different variants and the electric furnace process based on scrap. During the production of steel from pig iron the task is to remove the unwanted impurities dissolved in the iron – mainly C, Si and P – either via a slag phase or via the gaseous phase ("refining"). What all the reactions between the molten metal bath and the slag have in common is that oxygen is always involved. It distributes itself between the two phases according to certain equilibrium relationships, i.e. during oxidation of the unwanted accompanying elements part of the iron is also oxidized which in turn partly goes into the slag and partly dissolves in the steel bath. For the quantitative description of the reactions running during production primarily the law of mass action (C.M. Guldberg and P. Waage, 1867), Nernst's law of distribution (W. Nernst, 1891) and Henry's law (W. Henry, J. Dalton, 1803) are used.

Fig. 6.6.
Principle of the oxygen steelmaking process

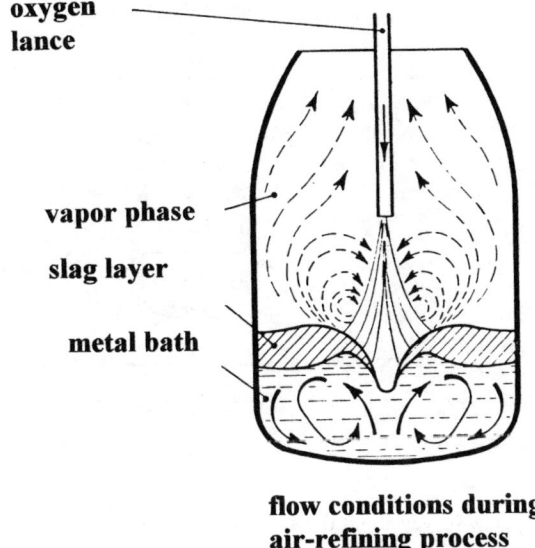

oxygen lance

vapor phase

slag layer

metal bath

flow conditions during air-refining process

The various procedures for steel production differ in the mode of adding the oxygen essential for the oxidation of the unwanted impurities. In the basic oxygen process mentioned first, pure oxygen (>99.5% O_2) obtained from air by means of the Linde process is blown onto the melt (Fig. 6.6). This process has meanwhile been extended by a number of variations in which, for example, inert gases or hydrocarbons are "blown" through the smelt together with oxygen.

The removal of carbon from the liquid iron (decarburization), which at the prevailing temperatures (about 1600 °C) leads to the formation of CO, is one of the most important refining reactions. The intense thorough mixing of the molten metal bath enables the fast transfer of matter of the reacting agents to each other and thus high melting rates. While carbon is removed in the gaseous state as CO, Si, P and Mn go, besides Fe, into the slag formed with the added lime. The various oxidation reactions provide the quantities of heat energy necessary for the procedure in the case of all basic oxygen processes.

The oxidation process in the case of basic oxygen steelmaking process only lasts 15–25 min, at the end of which rapid decisions for the further treatment of the melt have to be made. This short reaction time and the necessity of immediate instructions justifies the demand for an analysis "within a few minutes" or possibly without delay.

Because of its wide usage in present-day production, the following description will be restricted to the consideration of the oxygen steelmaking process (=L.D. process, named after the "Vereinigte Österreichische Eisen- und Stahl-werke" (VOEST) located in Linz and Donawitz, where the process was developed about 45 years ago), the oldest form of this steelworks technology. The L.D. processes in which the metallic charge can consist of up to about 30% of scrap

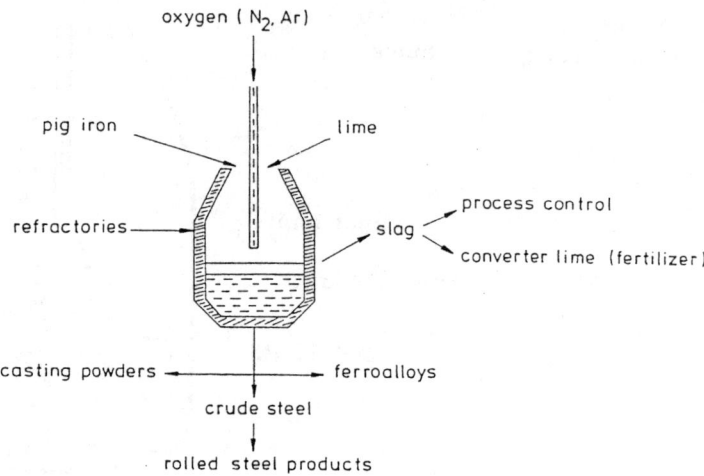

Fig. 6.7. Survey about the analytical fields connected with oxygen steelmaking

are characterized by high productivity. However, this advantage can only be made use of if at the right time the information (Information, analytically speaking: measurement and processing of the change in the state of distribution) required for controlling the process is available and can immediately be converted into decisions. The process control is supplemented by a production control system which calculates in advance from the orders received an optimum daily production, prepares up-to-date operations schedules several times a day on the basis of the particular actual situation and finally gives the process computers the scheduled data for production. Figure 6.7 gives a general idea of the wide range of materials connected with the converter process, a range which comprises all three states of aggregation and macro and trace analysis tasks and with which different analytical principles and time demands are linked. With regard to the refractory building materials which are used for the lining of the reaction vessel (converter) the problem of "speciation" can exist in the sense of the determination of the mineralogical composition.

The development lines of process analytics have always followed the demands of metallurgical process technology for speed and precision [12]. In particular, the basic oxygen technology described required the use of analytical methods performable within minutes and automatable for the primary product (hot metal), the reaction product (slag), and the final product (steel) in order to realize optimum operation conditions. This problem had been solved in particular by the application of atomic spectroscopic methods [13] (see Sect. 6.3.2.2) that can be looked at as a deciding prerequisite for the development of modern steelmaking processes [14]. The alloying and tramp elements to be determined within the scope of process control can vary in wide ranges of content. In certain

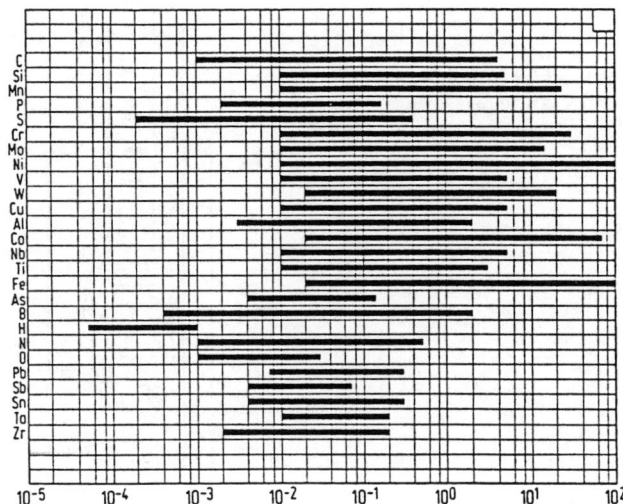

Fig. 6.8. Ranges of content to be analyzed in steel for process control purposes (in m/m%)

circumstances up to 30 chemical elements have to be determined on a routine basis over more than six orders of magnitude (Fig. 6.8).

In the case of unalloyed steels the determination of the carbon, manganese, phosphorus and sulfur contents is the primary task. This is followed by the elements silicon, aluminum, niobium, vanadium, titanium and nitrogen. However, the copper, chromium, nickel, tin, arsenic and boron contents can also be of interest for material-related reasons. In the alloyed steels molybdenum, tungsten and cobalt are added to the elements already mentioned. In special cases the determination of lead, antimony, zirconium, tantalum etc. can be necessary. With regard to the behavior of the steels in further treatment processes it is frequently necessary to know the oxygen or hydrogen content.

The assessment measure for successful converter operation is primarily a high "ability to achieve specifications setting accuracy" of chemical composition and temperature of the molten steel bath after the end of the process. In order to increase this ability, to avoid waiting times and for the quick removal of samples which are characteristic for the process stage reached, submerged lances have been developed which permit temperature measurement in and the taking of samples out of the upright standing converter [15]. Attention has already been drawn to the importance of sampling in line with the task and the material in the introductory chapter and made clear by example of the melt control in steel works (see Sect. 1.5).

Moreover, a carbon determination by means of an EMF submerged lance measurement of the oxygen activity or a determination of the liquidus temperature is possible [16] which in interaction with a dynamic process model can allow the advance calculation of the refining process end point. Such a multipurpose lance head which enables simultaneous carbon determination and

Fig. 6.9.
Section of a lance head for carbon
determination, temperature measuring
and steel sampling

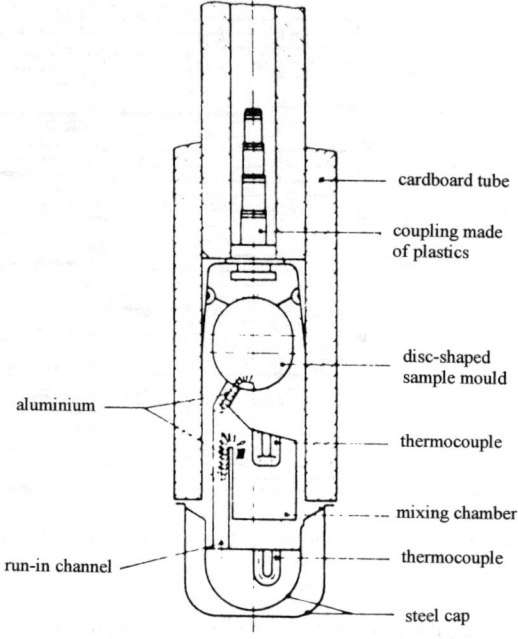

cardboard tube

coupling made
of plastics

disc-shaped
sample mould

aluminium

thermocouple

mixing chamber

run-in channel

thermocouple

steel cap

temperature measurement and at the same time provides a sample for spectral analysis is shown in Fig. 6.9. A previously widely applied control of the process is based on infrared spectrometric on-line analysis of the converter waste gases. For this purpose there are, moreover, proposals according to which mass spectrometry should be used for the analysis of all the components of the waste gases and for the increase of precision of these measurements.

Despite the successes achieved so far, analysts are faced with new demands as to increasing process automation and widespread application of secondary metallurgical measures [17]. A further increase in productivity could possibly be achieved, for example, by new modes of process analytics which deliver the information needed immediately and nearly without any delay. For the solution of this problem there are various research and development directions. One way is the complete automation of the well-known spectrometric analysis systems [6,18] so that from sampling or with the delivering of the samples in the melting shops all the process steps such as handling, sample preparation, sample transport to the spectrometer, analysis and data transfer to the plants proceed automatically (see Sect. 6.3.2.3) so that errors due to human shortcomings are ruled out [19].

Another approach consists of the installation of automatically functioning preparation and analytical equipment located in a container and set up in the immediate vicinity of the melting units. The charging of samples is in this case carried out manually by melting plant personnel (see Sect. 6.3.2.4).

Quite different from these intentions are projects for "direct" analysis of melts (in-line analysis), some proposals of which will be described in Sect. 6.3.2.5.

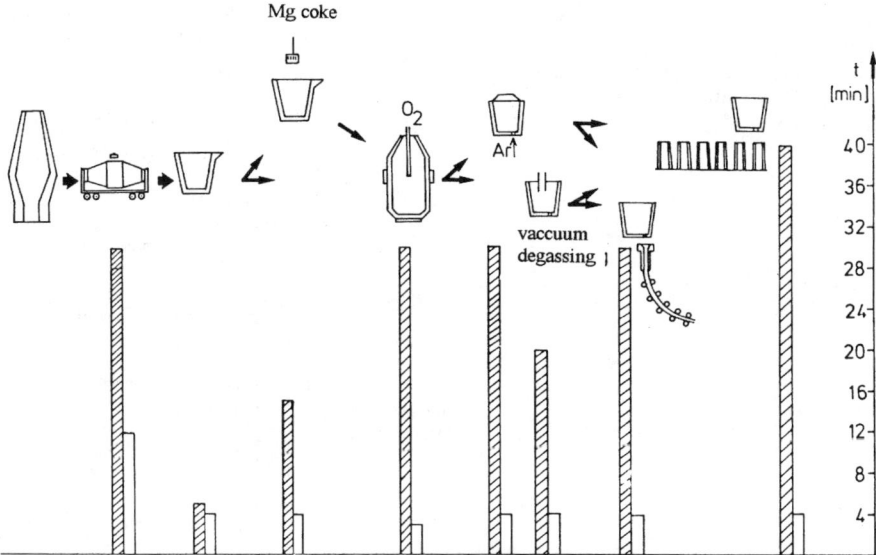

Fig. 6.10. Comparison of set-up times of the different metallurgical process steps with the corresponding times for analysis

In order to be able to assess the aim of these endeavours, a shortening of the "time for analysis", a glance should be taken at the development so far of spectrometric steel analytics: While in 1960 for the sequence sample handling, preparation, analysis and data transfer times of about 7–8 min were required, these times are today in the order of 2–3.5 min. Taking into account the plant start-up times these analysis times were previously in very many cases considered adequate (Fig. 6.10). If equipment for on-site analysis is used ("container laboratory") times for analysis of 1.5 min have been reached.

The chemical reactions taking place during steel production are temperature-dependent equilibrium reactions in the course of which every steel melt absorbs a certain quantity of oxygen, as between the molten metal bath and the slag there is, according to Nernst's law of distribution, an equilibrium relationship. If the melt is teemed off without further auxiliaries (i.e." unkilled") the major part of the oxygen escapes as carbon monoxide after reacting with the carbon. But not all steel grades can be teemed off "unkilled". At contents of, e.g., more than 0.15% m/m C and 0.50% m/m Mn there is no longer any adequate CO formation possible so that the oxygen has to be chemically bound by means of the addition of "deoxidizing agents" ("killed" teeming). From this point of view steel production is tied to the control of this oxygen movement and thus appropriate importance is attached to *oxygen analytics*.

As in the case of these reactions certain equilibria become established and a small amount of oxygen always remains in the steel. The deoxidation products formed (oxides) rise to the surface of the molten steel bath because they have a lower specific gravity than the steel and are insoluble in it. However, particular-

Fig. 6.11.
Immersion measuring probe for the
electrochemical oxygen determination
and the temperature measuring in
steel melts

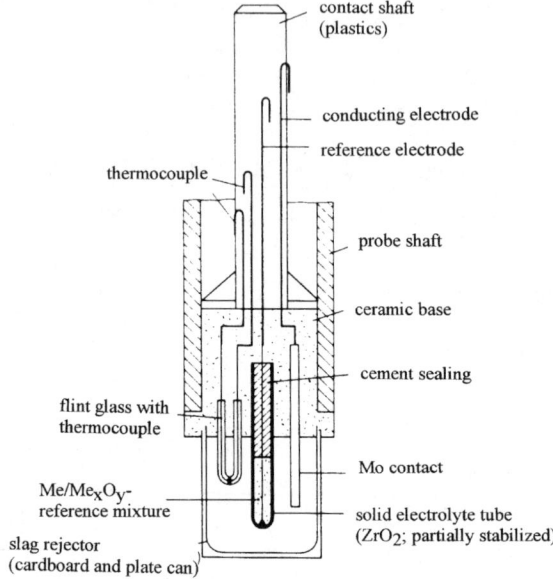

contact shaft
(plastics)

conducting electrode

reference electrode

thermocouple

probe shaft

ceramic base

cement sealing

flint glass with
thermocouple

Mo contact

Me/Me$_x$O$_y$-
reference mixture

solid electrolyte tube
(ZrO$_2$; partially stabilized)

slag rejector
(cardboard and plate can)

ly small particles remain in the steel melt and form – besides the abraded material coming from the refractory materials – "non-metallic inclusions" in the steel, the quantity of which can, in certain circumstances, be used for the characterization of the steel quality.

For the control and optimization of the addition of deoxidizing agents chemical analysis, for example with the help of fusing processes (fusing of the sample under a flow of inert carrier gas), was not suitable because of the great time expense and of the resulting amount of *total* oxygen. The consequent demand in this case for a rapid determination of the content of "dissolved" oxygen could be met by the development of electrochemical measuring cells [20]. Here, there are applied solid electrolytes, for example on the basis of stabilized ZrO$_2$ which have a predominant and high ion conductivity and an adequate resistance to thermal shocks [21]. The measuring duration according to this method is about 20 s, where the measurement also permits the identification of amounts in the range of some mg/kg. Figure 6.11 shows as an example a section through a disposable submerged measuring probe used for this purpose.

At this point it should be mentioned that work is in progress on the development of further electrochemical probes for process analytical in-line determination of accompanying elements in steel (in pig iron see above). Galvanic chains can be applied to the determination of the amounts of silicon (silicate electrolyte), phosphorus (CaO/CaF$_2$(Ca$_3$P$_2$)-electrolyte), chromium (slag electrolyte), aluminum (Al$_2$O$_3$-slag electrolyte) and sulfur (CaS-electrolyte).

The process step of deoxidization automatically provides a transition to the (further) secondary metallurgical processes which today are of decisive importance for the production of quality steels. At the start of this new process technology was the introduction into plants of the vacuum treatment of steel melts in the 1950s. For the removal of unwanted elements via the gaseous phase, besides carbon, the elements of hydrogen, oxygen and nitrogen can be considered as the reactions taking place here are pressure-dependent. Unlike hydrogen and nitrogen, the oxygen – as already shown – cannot be directly removed from the steel melt as a gas. It can only be separated from the bath together with existing carbon via the generation of CO.

Further secondary metallurgical measures concern the removal of unwanted accompanying elements, e.g. of sulfur, and the setting of narrow limits of the chemical composition of the steel melts. By the injection of solids (CaSi, CaC_2, Mg) and by the introduction of argon into the melt with the simultaneous formation of a reactive slag the sulfur content of pig iron or of steel melts can be considerably reduced and, thus, important conditions for an improvement of the steel quality can be created. By the use of secondary metallurgical procedures, total amounts of the accompanying elements C+S+P+O+N+H of 70 mg/kg can be reached, whereas in the case of hydrogen <2 mg/kg is the more likely figure.

From the ranges mentioned it follows that considerable demands are made on the analytics of the accompanying elements if the aim is in fact still to have the range 10–20 mg/kg safely under control for the individual elements. The problems are ruled by the homogeneity and the representativeness of the samples as well as by the influences of sample preparation, analysis and calibration (see Sect. 6.3.2).

Here, analytics are given a double role to play. On the one hand it is used for the characterization of the metallurgical effectiveness of a process, and on the other hand it is required in the form of a "rapid analysis" to describe the products obtained in the individual process stages.

While liquid steel in former times used to be exclusively cast as ingots after the setting of the required (chemical) composition which took place on the basis of the chemical analysis, since 1950 "continuous casting methods" have been increasingly applied. With this method the liquid steel is cast continuously in a water cooled mold in which it hardens apart from a still liquid core and then leaves the bottom of the mold red hot as a bar. The advantage as compared with ingot mold casting is that a number of work and handling procedures and the first forming step are dispensed with. Even here slight variations in the chemical composition can occur so that for supporting the quality control "quick" analytical examinations are necessary. That becomes all the more important when a series of several melts are cast successively (sequence casting).

The quasi-continuous casting of several steel melts suggests the idea of the development of processes for continuous steel production. Meanwhile there are world-wide endeavors and process technological approaches which mean that in the next few years greater industrial use can be expected.

6.3.2 Methods of Process Analytics

6.3.2.1 Sampling and Sample Preparation

The demands made by metallurgical process technology on process analytics have been described above and the analytical methods have been briefly mentioned. The sampling for the particular testing procedure and purpose represents the beginning of each test (analysis). Also for the production of metal materials great importance is attached to this step because of the demands made on the laboratory by the melting plant with regard to time and the quality requirements to be met by the plant. The sample to be taken from the melt must at the same time meet the qualitative requirements of the spectrometric analytical methods mainly used (e.g. freedom from blow holes and cracks; homogeneity) and be characteristic of the process stage reached (representativeness).

Sampling from the converters and other metallurgical vessels is done by dipping sampling probes into the melt, by which circular or oval test specimens (Fig. 6.12) are obtained which, after grinding or milling, are directly suitable for spectral analytical investigation. Only in special cases are steel spoons still used for sampling. The high level of sampling technology achieved after a development phase lasting several years [22] is reflected in the homogeneity of the samples taken by means of probes from the molten metal bath (Table 6.5). The samples are, in the event of a large distance from the analytical laboratory, conveyed by a fast pneumatic tube system (speeds of up to 30 m/s). In the case of the steel samples, sample preparation for the emission spectral analysis which follows is carried out by automatic milling or grinding machines for which many operating parameters can be specified depending on quality. The pig iron samples are also ground (see Sect. 6.2.2), while the slag samples, after being reduced in size, have a binder added to them and are pressed into tablets. More recent developments, because of various advantages, are heading in the direction of automated analysis systems [23] starting from automatic sample preparation and handling equipment (see Sect. 6.3.2.3).

Fig. 6.12. Samples for spectrochemical steel analysis

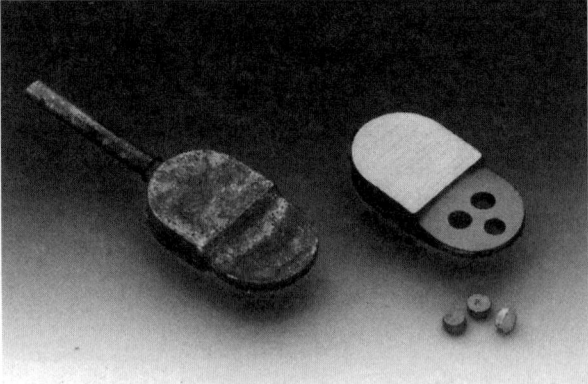

Table 6.5. Homogeneity of steel samples (data in % m/m) taken by sampler probes (example)[a]

CSi	Mn	P	S	Al	Cu	Cr	Ni
0.090	0.019	0.334	0.008	0.012	0.040	0.028	0.028	0.013
094	021	338	009	013	041	029	028	017
097	022	334	009	013	043	029	028	019
093	021	331	008	012	043	029	028	017
0.095	0.022	0.339	0.009	0.012	0.042	0.030	0.030	0.020
091	021	335	009	012	043	029	029	017
092	021	336	008	012	042	030	030	019
091	021	333	008	012	041	029	029	016

[a]Each side of the ground samples have been spectrometrically analyzed at four points

Fig. 6.13.
Influence of grain size on X-ray
intensity in the case of direct
analysis of powdery samples

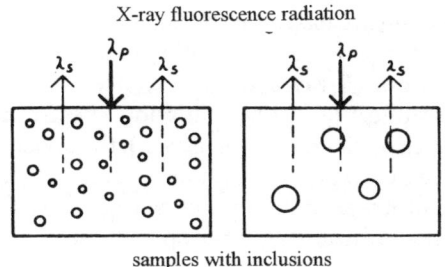

X-ray fluorescence radiation

samples with inclusions

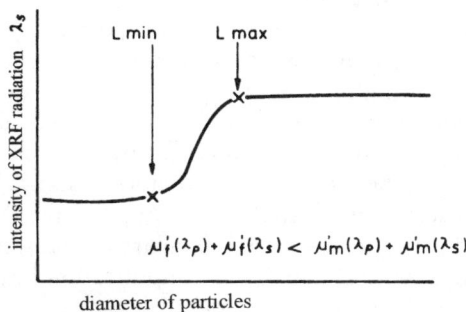

$$\mu_f^i(\lambda_p) + \mu_f^i(\lambda_s) < \mu_m^i(\lambda_p) + \mu_m^i(\lambda_s)$$

diameter of particles

Since the work step of sample preparation by grinding can have an influence on the precision and trueness of the analytical results, allowance must be made for this circumstance by means of quantifying preliminary investigations if a new analytical method is introduced. For example, the intensity of X-ray fluorescence radiation of some alloy elements depends to a considerable extent on the grinding conditions [24]. The influence of the microstructure in steel sam-

ples can, in certain circumstances, be eliminated by remelting with fast solidification of the melt (see Sect. 6.2.2). (The term "microstructure" comprises the
"entirety of observations with the naked eye or characteristics recognizable
under the microscope above all on the inside of a material".) With regard to
the X-ray fluorescence analysis of powdery samples it must be mentioned that
here a strong grain size influence exists (Fig. 6.13) which is eliminated by ultra-
fine grinding and the reproducible specification of the grinding conditions
[25].

6.3.2.2
Analytical Methods

The analysis of steel samples is performed, depending on requirements, on up to
30 elements preferably with optical emission spectrometry and that of pig iron
and of slags with X-ray fluorescence spectral analysis [26] (see Sect. 6.2.2), both
of which have reached a high technical degree regarding their reliability, speed
and reproducibility of measuring results [27]. The principles of this analytical
technique are based on the excitation, emission and optical dispersion of radiation of the wavelength of 150–600 nm (ultra violet and visible range) or
0.04–1 nm (X-ray and fluorescence radiation), respectively. With regard to the
applicability of optical emission spectrometry it is of importance that, for a
number of elements (e.g. C, S, P), the spectral lines especially suitable for analysis are below 200 nm and are thereby strongly absorbed by air. This difficulty
is eliminated by the sample being excited in an inert gas atmosphere (argon)
and the optical part of the spectrometer being evacuated (vacuum spectrometer).

In the case of optical emission spectrometry a certain quantity of the sample
to be analyzed is evaporated by means of an electric arc, spark or glow discharge
[28]. As a result of the high temperature of the plasma thus be formed, the atoms
of the vapour pass into an excited state, i.e. their electrons are brought up to levels of higher energy. Upon the return of these electrons to the basic state the
energy released is emitted as light, each element emitting a characteristic spectrum [29]. The light is spectrally dispersed by an optical grating or prism
(Fig. 6.14). For the determination of each element a suitable spectral line (light of
a certain wavelength) is collimated and its radiation intensity is measured. For
this purpose, special secondary electron multipliers are used which convert the
light with simultaneous amplification into electrical energy. During the evaluation of these electric measuring signals, in certain circumstances, inter-element
effects and/or the structural composition of the sample (matrix effects) are to be
taken into account. The measured values obtained do not, however directly represent the analytical results. Rather the allocation of the measured value indications to certain contents assumes the investigation of samples of known composition under the same conditions (calibration).

The time demands made on analytics have repeatedly been mentioned so that
at this point examples of the amount of time required for sample preparation
and spectral analysis should be considered (time in seconds).

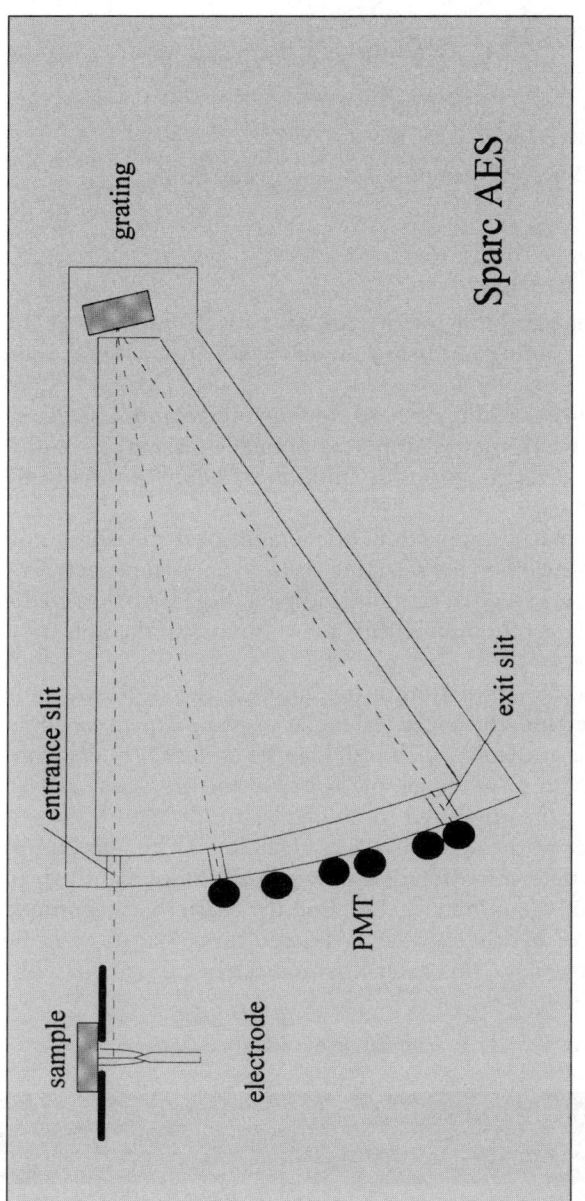

Fig. 6.14. Principle of optical emission spectrometry

1. *Sample preparation* incl.
   all handling and sample transport within the laboratory –60–80 s

2. *Spectral analysis* (duplicate determination)
   preflushing of the spark stand with Ar – 10 s × 2
   presparking (sample homogenization) – 10 s × 2
   integration – 8 s × 2
   data check and output – 10 s × 2

3. Total time – about 140–160 s

The time needed for slag analysis by X-ray fluorescence spectrometry is of the same order, when fusions becoming necessary requiring 30–60 s, while the net measuring time is merely 12 s.

Oxygen and hydrogen which are important in steel analytics cannot be determined by optical emission spectrometry. Moreover, optical emission spectrometry does not permit precise determinations of amounts of <0.01% m/m of carbon or <0.005% m/m of sulfur.

With regard to the systematic application of metallurgical measures and adherence to limiting values specified for reaching certain material properties – as already explained in Sect. 6.3.1 – great economic importance is attached to the analytical methods for the determination within a few minutes of the non-metals mentioned.

While in earlier times the determination of the "gases in the steel" was only possible with complicated and lengthy analytical methods using expensive apparatus, nowadays equipment is available which provides the required results within a few minutes [30]. For this purpose the sample is melted and the gases released are passed, after separation of the unwanted components, to a detector [31].

Table 6.6 gives a summary of the analytical-technical data of various equipment developments. For the determination of the oxygen and/or of the nitrogen content, both the principle of vacuum extraction and the carrier gas technique [32] are used. The analysis of hydrogen was considered for a long time to be problematic and difficult. Nowadays, this determination can be carried out with-

Table 6.6. Technical data of analyzers for the determination of carbon, sulphur, oxygen, nitrogen and hydrogen

Element	Measuring principle	Furnace temperature °C	Weighed sample g	Range of application μg/g	Time for analysis (1 sample) s
Carbon	IR[a]	ca. 2700	1	1–50,000	30–80
Sulphur	IR[a]	ca. 2700	1	1–4000	30–80
Oxygen	IR[a]	ca. 3000	0.5–3.0	0.1–2000	30–120
Nitrogen	HCD[b]	ca. 3000	0.5–3.0	0.1–5000	40–120
Hydrogen	HCD[b]	<2000	0.01–1.0	0.5–15	120–210

[a] Infrared absorption
[b] Heat conductivity detector

in a few minutes just like that of oxygen and nitrogen. In retrospect it can be stated that, in the main, analyzers working on the carrier gas principle have succeeded in becoming established.

At this point attention should be briefly drawn to a problem regarding the valuation of the hydrogen content determined. The meeting of a definite upper limiting value in the liquid steel does not always guarantee the content in the rolled steel product admissible for avoiding material faults. So far, however, too few systematic studies have been carried out on the change in the hydrogen content in the course of the various steel processing steps. The hydrogen can become concentrated at preferred places after solidification and in the worst case, in the absence of heat treatment or after incorrect heat treatment, lead to the destruction of the material.

6.3.2.3
Automatization of Process Analytical Methods in Central Laboratories

With the aim of reduction of the time for analysis, quality increase and the reduction of costs, extensive automation systems (system=demarcated arrangement of a quantity of elements [specified or selected smallest components] and a quantity of relations between these elements.) have been developed [33]. Figure 6.15 explains the principles of the lay-out and the various process stages of

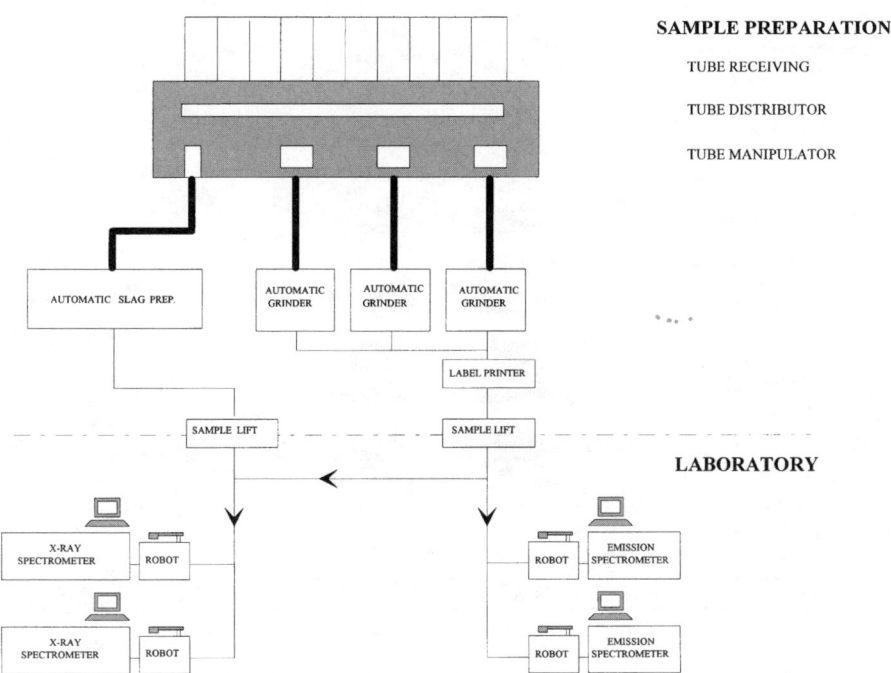

Fig. 6.15. Scheme of an automated steel works laboratory

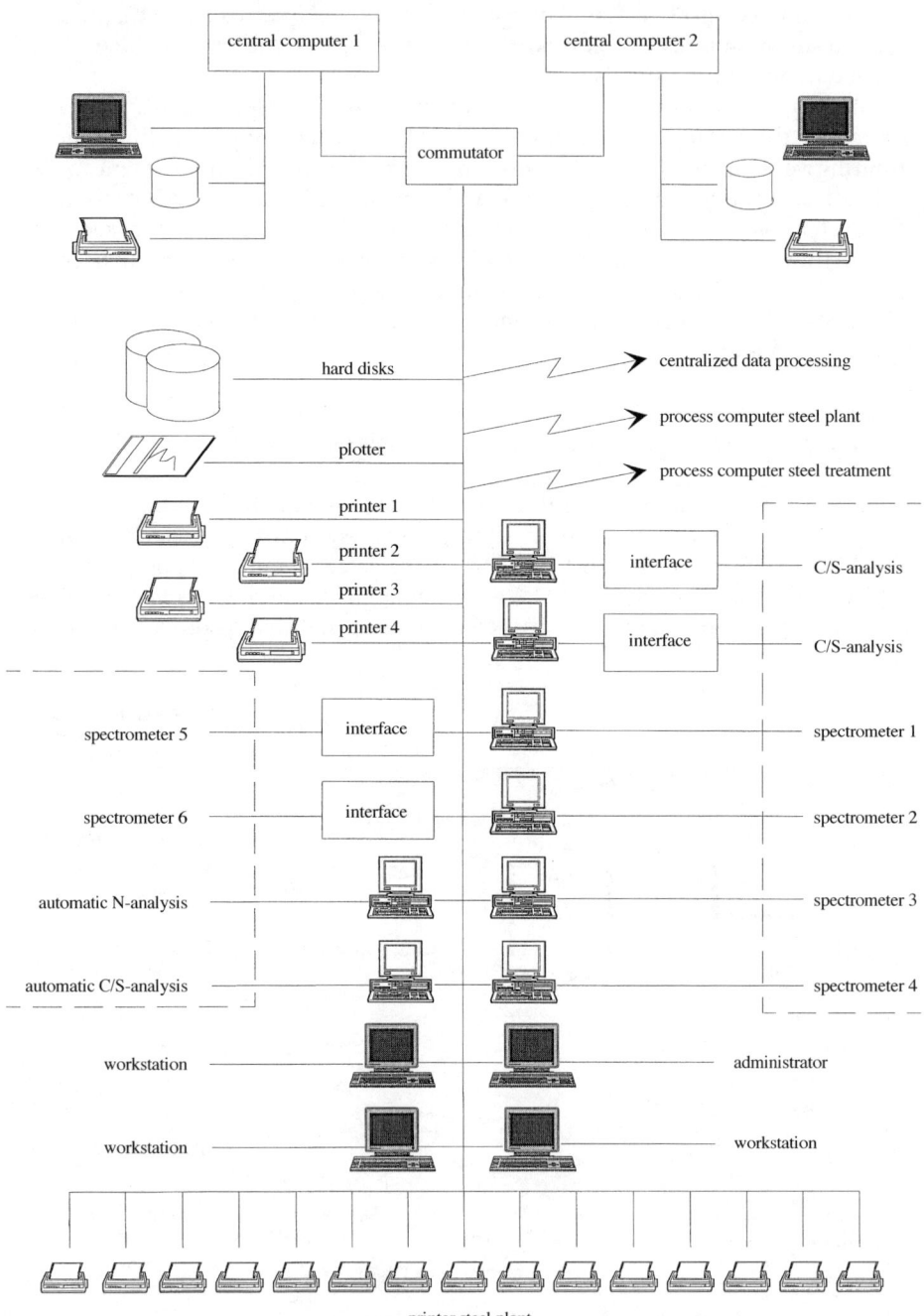

Fig. 6.16.A Configuration of an analytical data processing system using central computer installations

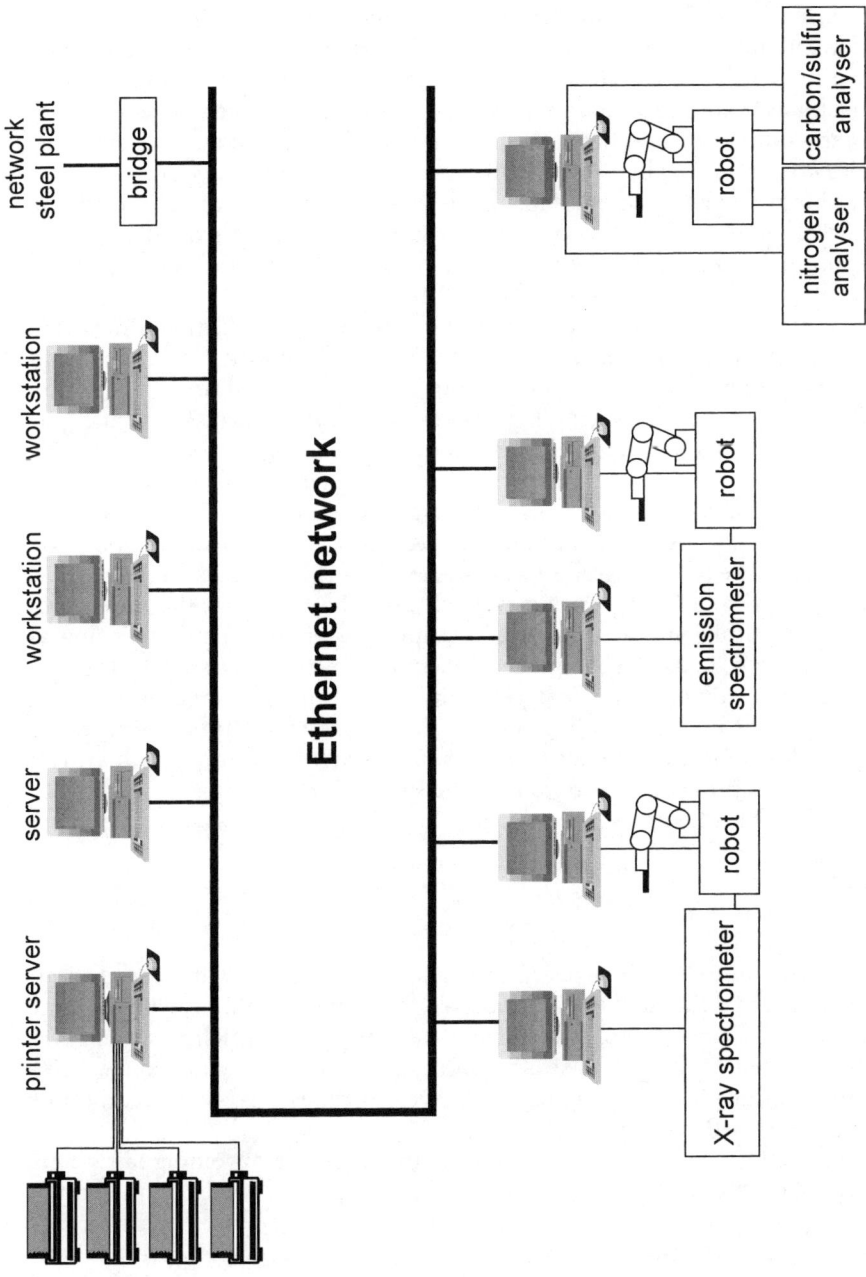

Fig. 6.16.B Configuration of a modern network-based data processing system

such a unit for metal and oxidic samples. Automation starts with the pneumatic tube sender and the receiving stations in the laboratory. This is followed by sample sorting which passes the samples priority-oriented to the automatic grinding machines or to slag preparation, with an automatic sample follow-up monitoring the total sequence. In the further course of the procedure the prepared samples are automatically passed to the spectrometers and analyzed.

For obtaining partial samples for the determination of nitrogen and oxygen, cylindrical samples are punched out of the sample body (approx. 1 g) and fed to the appropriate analyzers pneumatically. Here, too, there are initial solution approaches to the complete automation of these methods.

The carrying out of multielement methods with large devices and the rapid transfer of data to the melting plants and to the quality control departments naturally require an efficient data processing system in the laboratory and networking with a set over electronic data processing system [33]. As an example, Fig. 6.16 shows configurations realized in a steel works laboratory which explains the expenditure necessary for quality-oriented steel production.

For the correct use of large items of analytical equipment which are associated with large amounts of capital expenditure, aspects of the theory of operation [34] are also to be taken into account in order to meet the process technological demands with the lowest possible expenditure. That is to say, the number of items of analytical equipment or systems (capital expenditure) must be as low as possible without unacceptable queues for sample treatment arising for the customer or for customers. Therefore primarily the waiting times approved for the analyst are to be included in the considerations. At the same time priorities are to be laid down for the different types of samples [35]. In addition to this, there is the task of improving the economy of the system further by having batch jobs done during the "idle times".

6.3.2.4
On-Site Analytics

In connection with the increasing process automation and the widespread application of secondary metallurgical processes, ways were sought to avoid the time for transporting the samples to the analytical department which was, generally, centrally located. Even in the case of a laboratory located in the immediate vicinity of the melting plant, pneumatic tube conveying times of 20–30 s occur, a time which can be worth saving.

A first step towards shortening the time for analysis by avoiding fairly long transport times consisted of the installation of "shop laboratories" in the immediate vicinity of the melting units [36], e.g. on or under the converter platform of a basic oxygen or electric steel works. These laboratory sub-areas represent units supplementing existing central laboratories and contain the analytical equipment, spectrometers and individual element analyzers required for the whole of the process control, including all the auxiliary equipment for sample preparation. The personnel working in these laboratories is preferably placed under the head of the chemical laboratories and is stocked up by these central areas in the

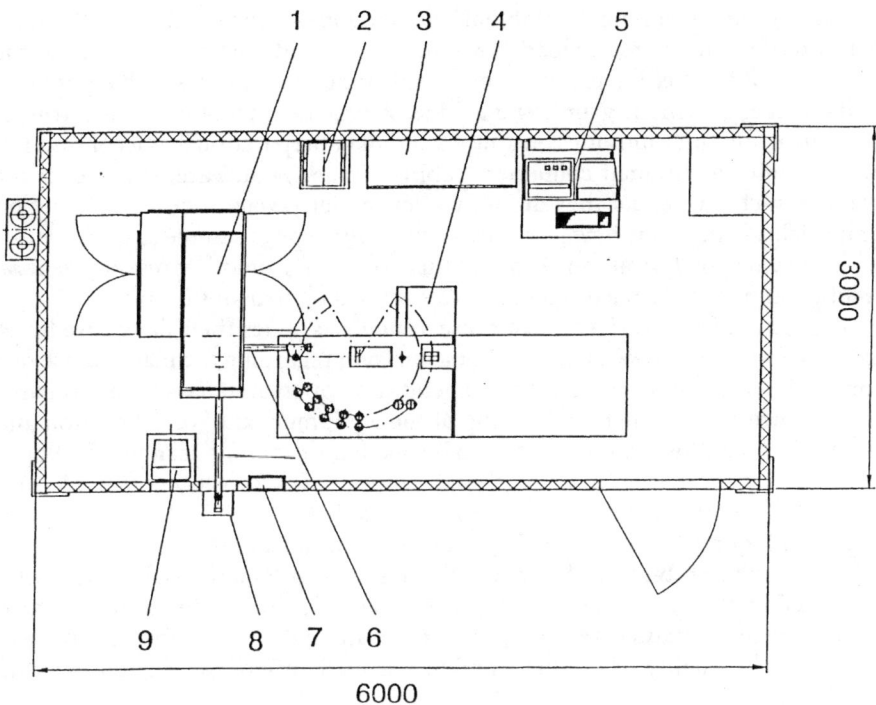

1 = Automatic grinding machine
2 = Voltage stabilizer
3 = Switch cupboard for grinding machine
4 = Inkjet printer
5 = Terminal

6 = Sample transport system
7 = Input keyboard
8 = Sample input position
9 = Input picture screen

Fig. 6.17. Example of a container laboratory

event of personnel shortages. As a result of constant need, and because of the maintenance and recalibration of the analytical equipment required to be carried out according to uniform criteria within the scope of quality assurance in line with standards, other organization forms can never be efficient.

The second step that followed was the automation of sample preparation both for the field of emission spectrometry and for that of the steel accompanying elements C, S, N, O and H, including their subsequent analysis [36], which enabled a marked reduction of personnel costs.

In contrast to shop laboratories in which *several* (automated) analytical lines, are present, e.g. for pig iron, for steel, and for slags, the operatorless plant container or cabin laboratories, as a rule, contain only one automatic analysis system. The more recent developments in equipment construction made possible the installation of spectrometers in the vicinity of the melting units (at-line)

[37–40]. Robust equipment combinations developed which withstand the temperature, dust and vibration loads near the process and can be "started up" by the personnel of the melting plant via the sample input. In these cases all equipment units such as automatic grinding machine, sample manipulator and spectrometer with a computer unit for controlling and data preparation are accommodated in an air-conditioned container (cabin) (Fig. 6.17) and are installed in the plant in such a way that they are only a few meters away from the production unit. Such a system can, however, only analyze one type of sample (e.g. only steel) with a limited analytical range so that this approach cannot represent a general solution and must be regarded only as a part of a total analytical task.

The functional effectiveness of these systems assumes their daily care by an analytical specialized department. For a decision regarding the planned use of a container laboratory, besides the expenditure on maintenance and repairs, expenditure on the functional testing of the equipment and for calibration and recalibration of the analyzer must also be taken into account. It should be borne in mind that the system must be shut down each day for routine maintenance work. During this time the samples must (be able to) be passed on to a second container laboratory possibly existing or to a central laboratory.

A particular problem in the case of the use of an automatic analysis system is presented by the identification of faulty samples (e.g. by slag inclusions or cavities). The faulty analyses resulting from the testing of such samples would lead to inadequate quality of products and to a disturbance of the sequence within the works.

Therefore, at an early stage methods for the prompt identification of faulty samples were looked for. One possibility consists of comparing the measuring signals of the spectrometric reference channels during sparking with desired data obtained from a sample free of faults. If the data are below these desired data the sample is then rejected. However, in the case of testing under operating conditions this method has proved insufficiently reliable. Another approach consists of the use of image analyzers which make it possible to identify defects in the surface of the samples and to sort unsuitable samples before spectral analysis. As samples containing slag inclusions and cavities (blow holes) are lighter than faultless, compact samples, as a third method a weighing procedure for the identification of faults and sorting can be used. This approach has already been successfully applied.

One aspect which deserves special mention is the marked rise in the percentage of samples without defects drawn from the molten metal bath if a container laboratory is used as compared with a conventional centrally organised laboratory operation. The explanation is the following: The skilled worker who has to carry out the sampling transports the sample not to a pneumatic tube system that leads to a laboratory which is "anonymous" for him, but he passes it himself into the analysis system located in the vicinity of process and thus he can directly control whether "his" sample can be analyzed. If the sample is not faultless it is rejected by the system, and the melter is thus compelled to obtain a new sample at once in order to limit the time delay. This direct feedback of good/bad information results in a drastic reduction of unusable samples.

6.3.2.5
In-Line Analytics

Fundamentally different research directions are taken by the work on "direct" analysis of melts (in-line analysis). The approaches used which have already been partly described in Sect. 6.2.2 relating to the direct analysis of pig iron are different.

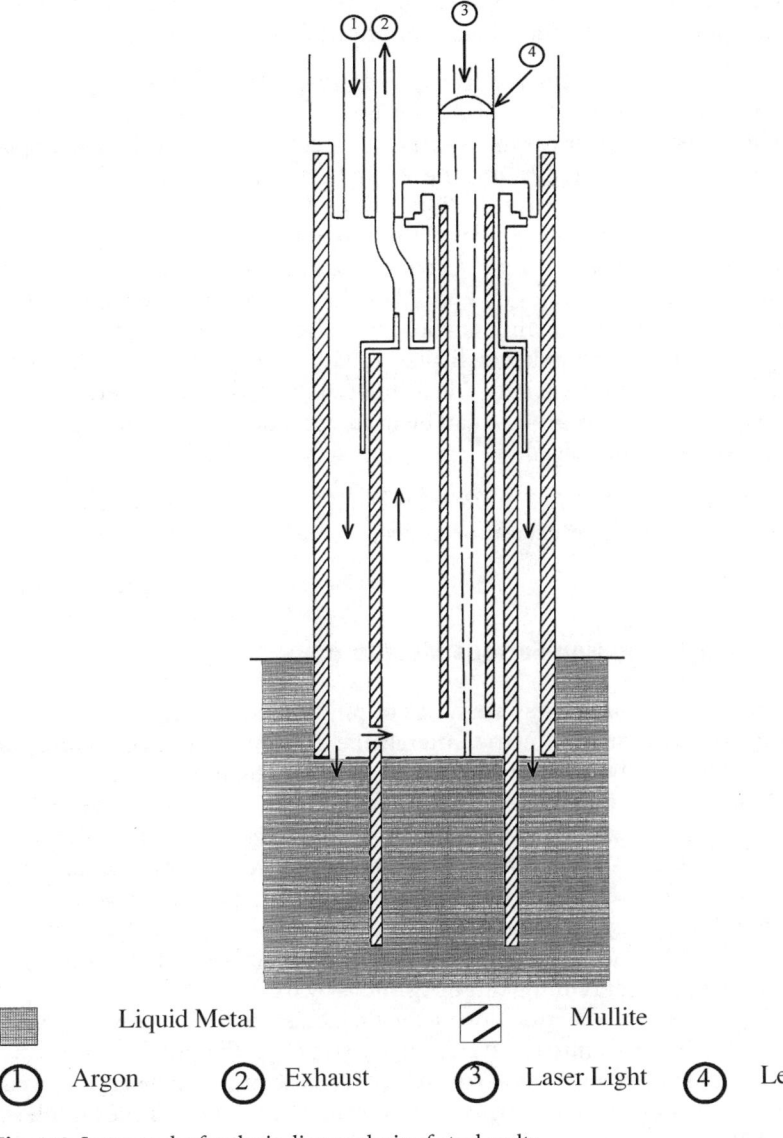

Fig. 6.18. Laser probe for the in-line analysis of steel melts

Whereas the thermal analysis proposed for "fast determination" of the carbon in steel (determination of the liquidus temperature) and the thermoelectric method required sampling from the melt, the EMF measurement of the oxygen activity (and thus the calculation of the C content) enabled for the first time a direct analysis of the molten metal bath. Here, a measurement of the potential between an atmospheric air (O_2)/zirconium dioxide electrode and the molten metal is performed. The EMF provides a measure for the partial pressure of the oxygen dissolved in the steel melt within some seconds [41].

Moreover, the following development directions make it possible to expect operational applications, besides the use of physical effects (ultrasonic).

1. *Application of a laser for the vaporization of sample material [42]*
   By this method, with the help of a laser beam, vaporized material is transported by an inert carrier gas to an emission spectrometer, which is equipped with a plasma source (ICP) for the excitation of these aerosol particles (Fig. 6.18). After testing of this technique on solid steel samples, investigations were started on the transfer of this method to liquid melts.
2. *Application of a laser [8, 9] or an electric spark as an optical excitation source*
   In this case, the spectral emission excited by the laser beam is used directly for spectrometric analysis. For this purpose, a cooled lance is inserted into the melting vessel, the slag blanket is displaced with an argon flow and the metal bath surface is spectrally excited. The initial tests on 30 kg melts were promising. However, before this process can be used operationally a large number of problems are yet to be solved.

Other researchers use openings in the vessel wall [43] or the vessel bottom [44] for the insertion of the laser excitation source.

6.4
Process Analytics in Non-Ferrous Metallurgy

The statements made on process analytics of pig iron and of steel also apply to the process analytics of non-ferrous metals, i.e. in non-ferrous metallurgy in principle the same analytical methods are used as in steel production.

Thus, *optical emission spectrometry* is of great importance for the metallurgy of aluminium and of copper (see also Sect. 4.3.3) as well as for the production of their alloys. It is also used in all the process stages of lead refining and alloying.

X-ray fluorescence spectrometry is used for analysing Cu, Ni, Zn and Al alloys, while, besides that, X-ray diffractometry for the control of fusion electrolysis of aluminum and of the bauxite decomposition process has been introduced [45].

The pyrometallurgical refining of Pb/Ag alloys by oxidation of the base metals requires a high degree of attention during the final phase of the process. By the use of *glow discharge optical emission spectrometry* (GDOES) the problem of precise process control has been solved and remarkable cost savings have been achieved.

Besides the widespread application of spectrometric methods, for example, in the zinc industry, titrimetric, photometric and electrochemical methods are used for process analytics. For the control of the leaching, cleaning and precipitating stages, on-line analyzers have been developed which can form a major part of an automation system in the course of process optimization. During the last few years X-ray fluorescence analysis is increasingly being introduced into zinc works.

The determination of oxygen and nitrogen within the scope of process control of non-ferrous metals, such as Mo, Zr, Cn, Ti, V, Al and W, is carried out, as described in the case of steel, either by means of vacuum or carrier gas hot extraction.

6.5
Analytical Investigations of Metal Forming and Annealing Processes for Sheet Production

The consideration of primary and secondary metallurgical process routes has shown that the extent, complexity and thus expenditure of process analytics accompanying them on the way to the final product are increasing. During the transition from liquid to solid steel the material is available in a form which in fact must still be subjected to the most varied forming and annealing processes but no longer in all cases requires process analytical control.

By far the greatest part of the crude steel produced in the steel works is processed in rolling mills. A small part is cast into molds and is thus used for the production of difficult (irregular) shaped parts, e.g. engine components, anchors (cast steel). While the pig iron and steel production is characterized by chemical reactions, rolling represents a physical process if one disregards the structural changes and phase transitions proceeding during the various heat treatments.

The "ingots" and "slabs" coming from the steel works, and as a rule still glowing, are, before processing, placed in furnaces in which temperature balancing between the inside of the ingot and the surface of the ingot and the necessary rolling temperature of 1200–1300 °C is carried out. The shape of the "ingots" depends on the further processing and thus on the type of finished product. Ingots for section steel and for steel bars have a square cross section, forging grade ingots are polyhedral prisms and seamless tubes are rolled from round ingots (cylinders). Ingots with a rectangular cross section are termed slabs. They are used as the base material for universal mill plate, plates and steel strips. The ingots' weights range from less than 1 tonne up to 400 tonnes in the case of forging grade ingots. During rolling the steel is exposed to high pressure between two rolling bodies rotating in opposite directions and thus is stretched – mainly in a longitudinal direction. The crystallites forming the microstructure are so markedly changed during the deforming of the rolled product that a considerable change in the physical properties occurs, strength and hardness increase, and ductility and toughness decrease.

At high rolling temperature (hot rolling), however, during or directly after rolling a **re**orientation of the crystallites (recrystallization) takes place as a result of which the strength increase occurring during deforming is annulled. During rolling in the cold state (cold rolling) the physical property changes remain. They are therefore to be followed by annealing treatment in order to adjust the required material properties. Further cold rolling (dressing) ensures a high surface quality. The widespread use of process computers enables a product-related control of the operating sequences and automation of rolling mills.

On the way from steel production to the production of final products by a system of procedures it is ensured (quality management system; see Chap. 10) that the material part made from a certain melt or from a certain prior product has a specified "quality". In such a system by "quality control" the degree is to be ascertained to which products (parts of products) match the demands made on them. By means of the methods of quality inspection, which mainly for time reasons must always be a "fast test", either mixing of material parts are to be excluded or deviations in the properties of the parts are to be ascertained in such a way that sorting according to quality characteristics is meaningfully possible.

The demand for simplicity and speed – as well as for the grinding spark test on steel materials used for more than 80 years – is met by magnetic and electromagnetic methods for the identification of various steel grades, which not least for these reasons and because of the non-destructive way of working [46] have been widely used during the last 40 years [47]. However, it must be noted that the heat treatment can exert a far greater influence on the electromagnetic properties than differences in the chemical composition. Therefore, these methods are applicable, for example, only if the microstructural state is known. Because of these limitations a search has been made for other possibilities of fast quality inspection, in particular chemical/analytical methods.

During the last few years a number of portable analyzers which used the principle of X-ray fluorescence have been developed for these purposes. They are subject, however, to certain restrictions and always only enable the determination of one or few elements. The excitation sources used, depending on the analytical task, are various radioisotopes for the determination of individual elements, both wave length dispersive and energy dispersive systems being applied.

In the field of alloyed steels these items of equipment can be used, by the defining of typical guiding elements, for tasks of material identification testing and for sorting purposes. For unalloyed steels no experience is so far available since in this case the applicability of this principle is even further restricted and the element carbon cannot be determined in this way.

The use of emission spectral analysis for product control tasks has a history going back more than 60 years ("Steeloscope" by Hilger, 1937). As the spark test mentioned above and the physical methods permit the identification of the chemical composition only indirectly and only under certain conditions, it was advisable to develop direct determination methods which, moreover, can be used at any operating points. In addition, the simultaneous determination of several elements was to be striven for, a specification which can only in some cases be fulfilled to a certain extent even by the more recent developed X-ray fluorescence analysers.

The last demands mentioned lead to the construction of portable or transportable spectrometers which have been applied in many cases in recent years.

With the help of an electric arc or spark between the piece of material to be tested and a counter-electrode, material is evaporated and in the well-known way spectrally excited.

The limitations indicated lead in the course of time to the following developments:

1. use of permanently installed (stationary) spectrometers in production lines (in-line analysis);
2. development of a mobile spectrometer linked to an aerosol generator;
3. development of mobile mini-spectrometers.

- *Stationary spectrometers for in-line analysis*

As a result of the development of fiber optical light guides it became possible to separate spatially the spark stand and the spectrometer. After the overcoming of numerous difficulties regarding the trouble-free installation of an emission spectrometer directly in a production plant, the providing of suitable spark testing equipment and the determination of the excitation parameters useable for direct analysis the conditions for the grade testing in a production line were created (in-line control) [47].

The testing stand comprises a movable milling device for producing a suitable test surface (front end of a steel bar), the movable spark stand, which is pivoted in front of the product to be tested, and the light guide which leads the emitted light to the spectrometer. The recalibration of the system is carried out automatically. The spectral analysis takes 5 s (presparking 2 s, integration 3 s). The relative standard deviation amounts to 2–8%. The cycle sequence time is 60–90 s (depending on the product to be tested).

- *Direct analysis with the help of an aerosol generator*

By this technique it is not, as above, the emitted light but an aerosol obtained from the product to be tested which is passed to the spectrometer. In this case, with a direct current arc, material is evaporated and fed as an aerosol in an argon flow to a capillary arc for optical excitation. The advantage of this method is to be found in the fact that all the elements determined on compact samples by the usual emission spectrometry, that is the important steel accompanying elements carbon, phosphorus and sulfur, are covered. However, this method has not succeeded in becoming very widespread.

- *Mobile mini-spectrometers*

The more recent development of equipment for spectral analytical identity testing is characterized by the construction of mobile mini-spectrometers in conjunction with a "sparking pistol" and transmission of the emitted light by means of a light guide [48, 49]. The measuring procedure for simultaneous determination of, for example, ten elements takes 4–10 s. The deviation of a measuring value from a specified content limit can be signalled optically and acoustically. Table 6.7 shows some typical examples of application.

Table 6.7. Examples for spectrometric identity tests of steels

No.	Steel grade	Element used for characterization	Mean value % m/m	Further possibilities for characterization
1	20 MnCr5	Chromium	1.25	1.20% m/m Mn
	RSt 37–2		0.11	0.70% m/mMn
2	St 52–3	Manganese	1.28	–
	St 37		0.62	–
3	42 CrMo4	Manganese	0.70	0.18% m/m Mo
	16 MnCr5		1.20	0.02% m/m Mo

Because of their importance for the consumption goods industry, cold rolling and the associated annealing treatments for sheet production should be briefly dealt with reference to process and product analytics.

Progressive automation of the manufacturing technique in the sheet fabricating industry is leading to increasingly greater demands being made on the homogeneity of the plate material properties. Besides the properties of the basic materials, in the case of metal sheets and black plates to an increasing extent the surface properties have gained in importance. In automotive manufacturing, for example, with a view to achieving perfect lacquering the highest demands are made on the surface quality as to absence from defects, roughness and cleanness.

The resultant demands made on the material are fulfilled by means of costly measures at the various stages of production which start with the pickling (removal of the scaling layer by chemical or electrochemical procedures in order to achieve a clean surface) of the base material. After cold rolling there are various possibilities of annealing treatment. The new continuous annealing processes, in comparison with the batch-type annealing furnace, offer, as a result of the flexible annealing conditions with annealing temperatures of up to 850 °C, ideal conditions for the implementation of the new material concepts adjusted to today's market requirements.

In the continuous annealing units process steps such as strip cleaning, temper rolling (rerolling of cold rolled plates and strips after annealing for leveling and strengthening of the surface) and electrostatic oiling are included. The special advantages of "continuously annealed" material to be emphasized are the good flatness, the strip cleanness and the suitability for surface treatment.

Continuously operating measuring equipment which is based on the *X-ray diffraction principle* and used for on-line control of an "anisotropy index" [50] and for detecting surface faults serve the quality control of the product. In the case of cold rolled plates the deforming properties are naturally of particular interest. The crystallographic fine structure of the plates (texture) can be seen as a physical parameter to be assigned.

The orientation of the crystallites within a macroscopic solid can range between two extreme cases, a disorderly distribution and strict order. The entirety of the orientation distribution of the crystallites is termed texture. Textures measured by radiographs can, for example, be described with the help of a

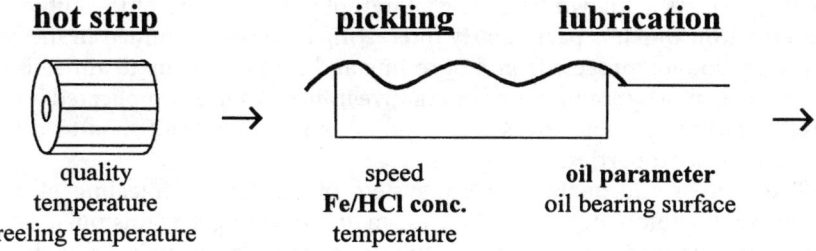

hot strip

quality
temperature
reeling temperature

pickling

speed
Fe/HCl conc.
temperature

lubrication

oil parameter
oil bearing surface

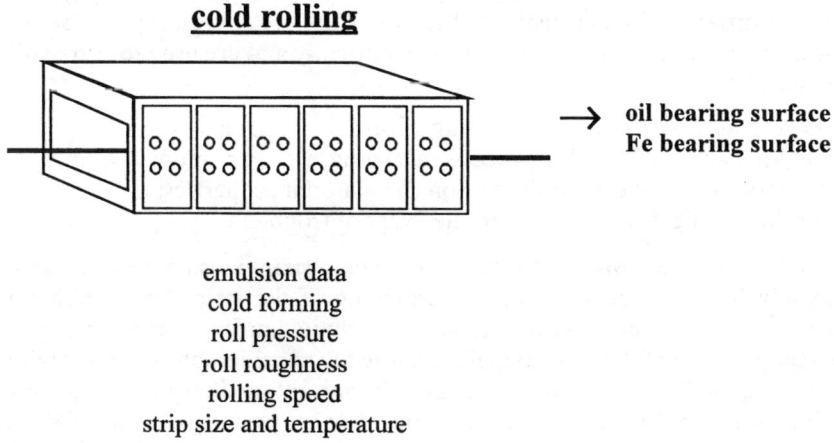

cold rolling

→ oil bearing surface
Fe bearing surface

emulsion data
cold forming
roll pressure
roll roughness
rolling speed
strip size and temperature

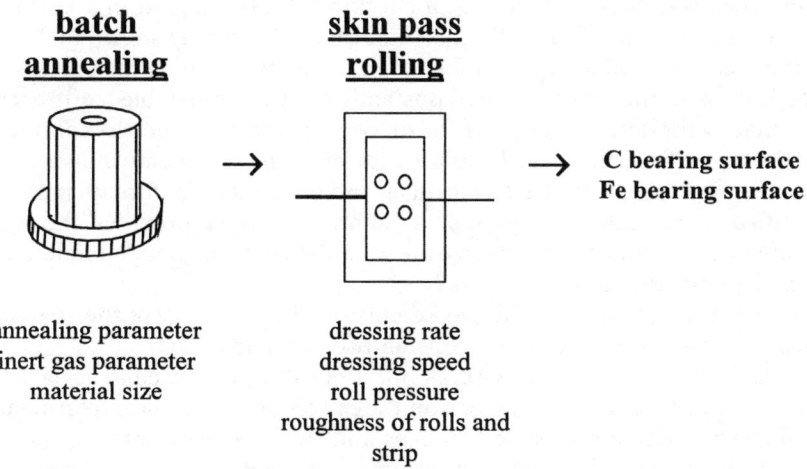

**batch
annealing**

annealing parameter
inert gas parameter
material size

**skin pass
rolling**

dressing rate
dressing speed
roll pressure
roughness of rolls and
strip

→ C bearing surface
Fe bearing surface

Fig. 6.19. Analytical control of the cold-rolling process "pole figure" as is known from crystallography. The texture of the cubic body-centered iron which is particularly interesting here is determined in the course of the production process from hot rolling and cold rolling up to annealing and temper dressing by a number of interactive influences in a complicated way. Different heating-up parameters lead to different textures and thus to different technological properties.

The process and quality control system of a modern annealing plant also comprises an automated laboratory for material testing. For this purpose samples from current production are regularly taken from the strip. After the automatic production of standardized test specimens the roughness, hardness (according to Vickers) and the tensile strength (tearing test) are determined.

Chemical process analytics (Fig. 6.19) of the cold rolling and annealing processes shows different characteristics to the case of the metallurgical processes in the liquid phase. All tests are carried out intermittently and are used for three different aims:

1. as process analytics in the meaning of process control;
2. for the "analysis" of the process regarding a better understanding of the process steps and their influence on the material properties;
3. for the clarification of causes in the event of trouble.

In the last few years the demands made on cold strip lines have increased considerably. Thus the greater overall deforming of the strip with simultaneous expansion of the production range to include harder grades and greater throughput rates of the cold strip lines has led to considerably greater strains on the rolling auxiliary materials. The most important auxiliary material for cold rolling is the rolling oil emulsion which has to perform several tasks (see Sect. 3.2.1.3). Its properties have a considerable influence on the "cleanness" of the product metal sheet. With the increased demands made on the cold strip lines, in order to achieve good strip cleanness without hampering the following treatment processes, increased analytical checking was demanded in addition to a constant adjustment of the rolling oil emulsions. As already shown in Table 3.5, rolling oils consist of a large number of ingredients, of which the presence and effectiveness in the emulsion are constantly tested. From Table 3.6 it becomes clear that by the determination of the oil content, the pH value, the conductivity and the saponification value, the lubrication efficiency of an emulsion is ensured just as foreign oil ingress, bacterial attack and penetration of foreign ions can be identified analytically. The use of a laboratory robot provides the already explained possibility for efficiently meeting the rising demands made on analytics (without additional personnel).

Besides the state of the rolling oil emulsion, the efficiency of the strip cleaning, pickling and oiling units are constantly controlled analytically.

Of major interest are, of course, all questions that are related to surface properties which are of direct influence on the procedures of surface treatment. For rapid characterization of sheet surfaces and their changes during processing, glow discharge optical emission spectrometry (GDOES) has proved very useful. The determination of the surface enrichments, for example when being annealed

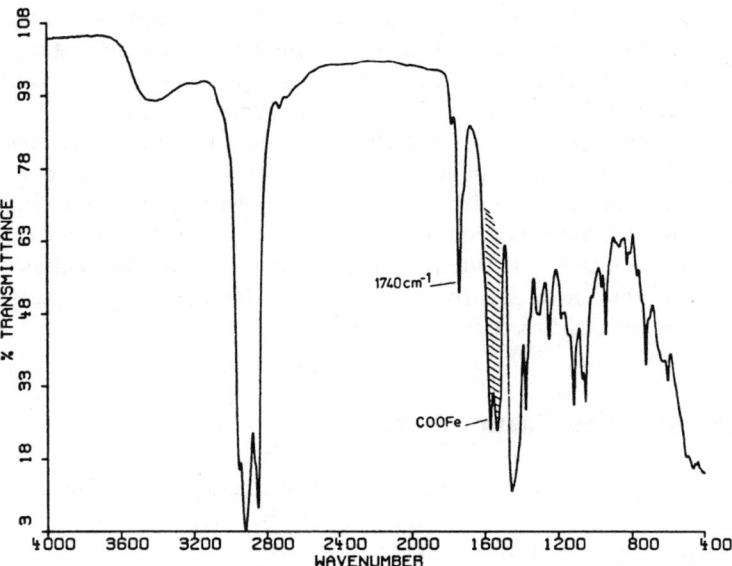

Fig. 6.20. IR spectrum of annealing residues on metal sheet

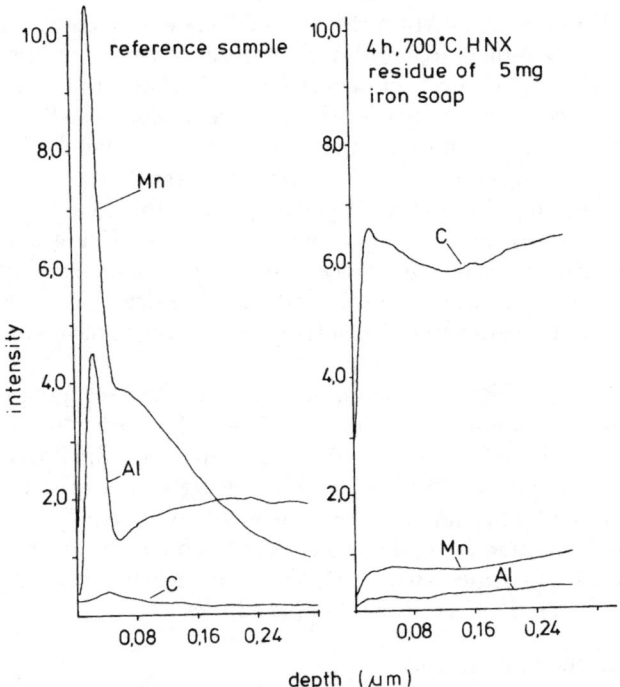

Fig. 6.21. Surface carbon be formed by annealing (GDOES depth profile)

under protective gas, is important for obtaining information on possible detrimental effects on following process steps (for further details see Sect. 6.7.3).

By infrared and mass spectrometry it was shown, among other things, that during the cold rolling process iron soaps are formed by reaction of rolling oil components with the steel surface. After annealing these carboxylates lead to residual carbon deposits which cannot be removed. Figure 6.20 shows an infrared spectrum of the residue of iron carboxylates after annealing which, even up to 650 °C, reveals organic constituents such as esters and carboxylic acids while the GDOES depth profiles in Fig. 6.21 provide evidence of the increase of residual surface carbon as a result of the annealing of plate samples contaminated with carboxylates.

6.6
Control of Surface Coating Processes

The coating of materials and the production of layers on materials have gained outstanding technical and commercial importance. There are a large variety of processes for sheet coating with metals (Zn, Sn, Pb, Al) and with organic substances in the form of foils and plastisols. Examples of ceramic or oxidic coating processes are enameling, the application of ceramic layers on turbine blades, or of ruthenium oxide on titanium anodes for chlorine alkali electrolysis for the extension of their service lives. Moreover, in this connection, aluminum electrolytic oxidation must be mentioned. The surfaces of objects made of iron and steel are frequently subject to, besides mechanical (erosion) influences, marked chemical and electrochemical influences (corrosion) and must be protected in such cases. An exception is only made by a few corrosion-resistant steel grades.

Steel sheet and black plate are often surface treated (coated) in order to prevent corrosion and, in certain circumstances, to improve the processability in the following process steps by coating of metal layers. In this case, it is a question of steel strip processing in continuous procedures comprising several process stages and existing in various process engineering variants. The individual procedures and their physical/chemical and metallurgical bases are not to be described here in more detail; rather it is only necessary to have a look at process analytical aspects.

The black plate (thickness<0.5 mm) intended for the packaging sector is electrolytically tinned ("tinplate") or chromium-plated in order to avoid corrosion phenomena. Continuous monitoring of the tin layer thickness during the process [51] represents a major means of ensuring a high product quality and cost minimization by the saving of tin. The measuring principle consists of the weakening of the excited $Fe-K_{\alpha}$ fluorescence radiation by the Sn layer lying above it. Different Fe/Sn_2 intermediate layers do not influence the measuring result.

For sheet coating there are different possibilities:

- hot dip and alloy galvanizing;
- electrolytic coating with zinc, zinc/cobalt, zinc/nickel and zinc/iron layers, and mixed lead coating.

The automotive, structural, and the apparatus industries are among the most important customers of these strip coated steel products.

In contrast to the hot dip galvanizing processes for the sheet coating with zinc or zinc/aluminum alloys ("Galfan"®:95% m/m Zn, 5% m/m Al; "Galvalume"®: 55% m/m Al, 43.5% m/m Zn, 1.5% m/m Si) which mainly need off-line control of the molten metal and the raw materials and by technological values, the electrolytic processes require a more comprehensive process control including process analytics.

The process analytical tasks start with the examination of the degreasing and pickling baths, then cover that of the electrolyte and of the rinsing, phosphating and passivation baths. The analyses are carried out off-line, mainly titrimetrically. There are various on-line concepts in this field using spectroscopic [52] and chromatographic principles [53].

In the literature there are already proposals for solutions using X-ray fluorescence analysis [54] and ICP spectrometry [55] which at least enable one to provide a considerable amount of information required for process control within a few minutes, but it is not a comprehensive solution to the problem.

For investigations of the metallic coatings GDOES and SNMS have proved to be widely applicable analytical methods (see Sect. 6.7.3).

For the sake of completeness the non-metallic inorganic coatings should be mentioned. In that field, enamel has gained particular importance. They are mostly vitreous alkali borosilicate layers which are coated and "fired" on the steel surface in one step or several work cycles (depending on the type, 700–1000 °C; enamelling). In addition, protective layers of cement (for example for the internal lining of pipes) and pure SiO_2 are known. These technical processes are also linked with numerous analytical tasks. The optimization of this technology succeeded only by systematic process analytical investigations.

A wide field is surface protection by means of paint coats with colours, lacquers, bitumen and plastics. Recently (since about 1960) plastic-coated steel has become increasingly important. In the case of this material the good mechanical properties of steel are combined with the corrosion resistance and other desirable properties of the plastics. These organic protective layers are applied to the steel strip in continuously working plants either as a liquid (plastisol) which is "hardened" under heat or in the form of foils. The plastics used for coil coating are polyesters, acrylates, epoxides, melamines, silicon polyesters and also polyvinyl chloride and polyvinyl fluoride. In order to ensure good adhesion between steel and plastic, as a rule, special treatment of the metal surface (e.g. phosphating and/or galvanizing) is necessary before coating. In addition, there is the process of lacquering with zinc dust (Zincrometal®; Zincal Duplex®) of cold rolled steel strip.

The coil coating mentioned has been one of the most important surface protecting techniques in the steel industry for more than 30 years. Coil-coated flat steel products can be used to advantage in cases in which corrosion resistance and decorative appearance are of decisive significance, e.g. in the structural industry and in the automotive, EBM, and packaging industries.

The coil coating processes again demonstrates impressively that, starting with the continuous annealing of steel strips, a process chain of continuous treatments in steel processing can lead to an almost unlimited variety of products depending on the demands made on decorative appearance, durability and formability.

The multilayer structure of such products brings about a situation in which a number of problems can occur on the phase boundary faces, for example in connection with questions of adhesion. Even if process control is primarily based on the measuring of physical values, analytical investigations are necessary to create the basis for improving the products concerned, as well as for the optimization of processes, and for producing data for improved operational control. The product range is supplemented by three-layer sound-insulating composite plates of steel/plastic/steel (Bondal®) for which, with regard to the analytical product control, the same applies as for the other multilayered products.

6.7
Analytics of Metallic Materials

6.7.1
Material Properties and Trace Analysis of Steel Materials

Several times reference has been made to the connection between the development of materials suitable for the market and the tasks, automatically related with it, of material testing and chemical analytics which frequently leads to process analytical methods fit for plant management. In this case, it is the task of "destructive" and "non-destructive" material testing, to determine the properties of the materials produced with regard to their practical applicability. By destructive material testing, for example, strength and hardness tests on special test specimens taken from the material are to be understood. In addition, the investigation of the forming of the microstructure (metallography) is performed by light and electron microscopy. Non-destructive testing is applied, e.g. for the determination of inner defects in workpieces using ultrasonic or radiographic methods for which as radiation sources X-ray tubes or radioactive preparations, and in the case of great thicknesses betatron or linear accelerators, are used.

Trace analytical investigations are necessary due to the fact that trace amounts of certain elements in steel (range 10^{-3} to 10^{-4} % m/m) have a decisive influence on mechanical properties. Here, too, reference is made to the fact that the information provided must be regarded as examples. As possible traces in steels the following elements can be considered (italics means used as micro alloy element): *Al*, As, *B*, Bi, Ca, *Ce*, Co, Cr, Cu, Mg, *Nb*, Ni, Pb, Sb, Se, Si, Sn, *Ta*, Te, *Ti*, *V*, Zn, *Zr*. The ranges of contents of the desired or undesirable elements are shown in Table 6.8.

There may be different reasons for the metallurgical and metallographic behaviour of trace elements (Table 6.9). A distinction is to be made between the immiscibility of the components in the solid state (Pb, Bi, Te, Ca), mixed crystal formations (Mn, B, Cu) and the formation of intermediary phases (As, Sb, Sn). The effects observed can be summarized under the following concepts, and at

Table 6.8.
Examples for the order of
magnitude of trace ele-
ments in steel

Element	% m/m
As	0.005
B	0.0005
Bi	0.0005
Co	0.004
Cu	0.08
H	0.0003
Mo	0.01
Ni	0.04
P	0.015
Pb	0.002
S	0.015
Sb	0.002
Sn	0.008

Table 6.9. Effects of some trace elements on typical steel properties

Element	Negative effect as tramp element	Effect as alloying element
Carbon		Hardness, tensile strength, ductility
Sulphur	Red shortness, hot shortness	Cutting property
Phosphorus	Brittle fracture, temper brittleness	
Nitrogen	Embrittlement (age-hardening), blue brittleness	
Oxygen	Deterioration of ductility and endurance strength	
Hydrogen	Crack sensitivity, flake formation, brittle fracture	
Boron		Hardenability, aging resistance
Lead	Tensile strength	Cutting property
Tin	Impact strength	
Antimony	Hot forming property	
Bismuth	Tensile strength, hot forming property, hardenability	

the same time the significance of trace analytics for the production of steel materials can be derived:

- influence of diffusion processes;
- influence on formation of crystal nuclei in the liquid phase;
- formation of low melting grain boundary eutectics;
- structural and energetic influencing of grain boundaries.

The unwanted trace elements come into the steel during the various process stages from the raw, auxiliary, and alloy materials. Therefore chemical analysis is essential for the selection of these materials used (e.g. iron ores, ferro-alloys). It can easily be seen from the explanations above that lowering of the trace and tramp elements will be a permanent aim of the metallurgists. Today 10–20 mg/kg of the non-metals in steel are attainable by special treatments. These figures make it clear that further metallurgical progress is not only closely linked with the development of analytical methods but also that this progress depends directly on it. Even in the case of the "super alloys" the trace elements play an important part. In this case, the limiting contents of detrimental "trace elements" are of the order of 0.5 (Bi) to 50 mg/kg (Sn, Zn, Cd), the attaining of which and following of which in certain circumstances already requires a limitation of the harmful elements in base materials to of the order of 1 mg/kg and less.

The different methods of trace analytics cannot be dealt with in more detail at this point. It can only be mentioned that for these tasks in the field of iron and steel, mainly emission spectrometry with spark and plasma excitation sources [56], X-ray fluorescence spectrometry [57], atomic absorption spectrometry in conjunction with flameless excitation techniques [58] and recently the ICP and GD mass spectrometry (inductively coupled plasma – glow discharge mass spectrometry) [59] are used. Determination limits of 1–0.1 mg/kg are thus achievable by appropriate choice of method. However, the problems of calibration and the availability of reference materials are posed with corresponding acuteness, more so in the case of trace and ultra-trace analysis in the ceramic and microelectronic industries [60–63] such as in the investigation of purest aluminum [64]. Interestingly, electrochemical methods have not established themselves with lasting success in this field of analytics.

6.7.2
Phase and Microanalytics

Like trace analytics, phase analytics also has a strong process analytical component. As a result of the thermal and thermomechanical treatments of steel the required material properties are finally obtained. Thereby effects are made use of which are based on the formation of dissolved or precipitated nitrides, carbides, and carbonitrides. The clarifying of relations between process parameters and metallographic alterations enables the formulation of regulation models and codes of practice.

A further interesting analytical task consists of characterizing inner boundary surfaces with regard to their foreign atom covering and determining their influence on the material properties. With these boundary surface layers, deposits of about 1–1.5 atom layers (0.5 nm) are involved. In polycrystalline materials the inner boundary layers can be described by crystallographic and chemical criteria. At the connecting planes between the crystals, lattice defects frequently occur. For the investigation of these boundary surfaces scanning electron microscopy (SEM) [65], Auger electron spectroscopy (AES) [66], transmis-

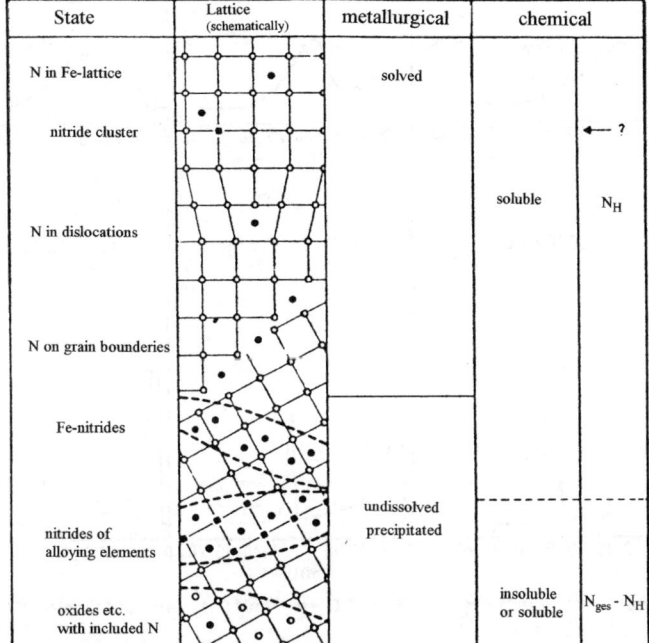

Fig. 6.22. Nitrogen phases according to Swinburn

sion electron microscopy (TEM) [67] and, widely used in other fields of instrumental microanalysis, electron beam microanalysis (EBMA) [68] are used.

For quick information on metallographic alterations during annealing treatment, investigations of phase and microanalysis of nitrogen in steel are currently seeing a renaissance. Nitrogen occurs in steel in very different states (Fig. 6.22) [69]. Its phase analytical determination is correspondingly complex. Besides the electron spectroscopic methods regarding rapid information, the proposal for the measurement and interpretation of evolograms obtained by temperature programmed heating of steel samples is of justified interest [70]. Furthermore, the hydrogen reduction method should be mentioned in this context [71]. This method will probably remain restricted to the determination of one nitride or of few nitrides of the same order existing in parallel. In the case of the latter method, steel chips are heated to temperatures of about 400–450 °C in a H_2 flow and the NH_3 generated by reduction of the uncombined nitrogen and of iron nitride is registered as a function of time (Fig. 6.23). From interpretation of the extraction curve, data on the degree of the nitride formation can be derived.

If it were possible to develop a rapid method for routine control on this basis it would be the beginning of a new era of process analytics: It would be the first example of a method in minutes for *process accompanying phase analysis*.

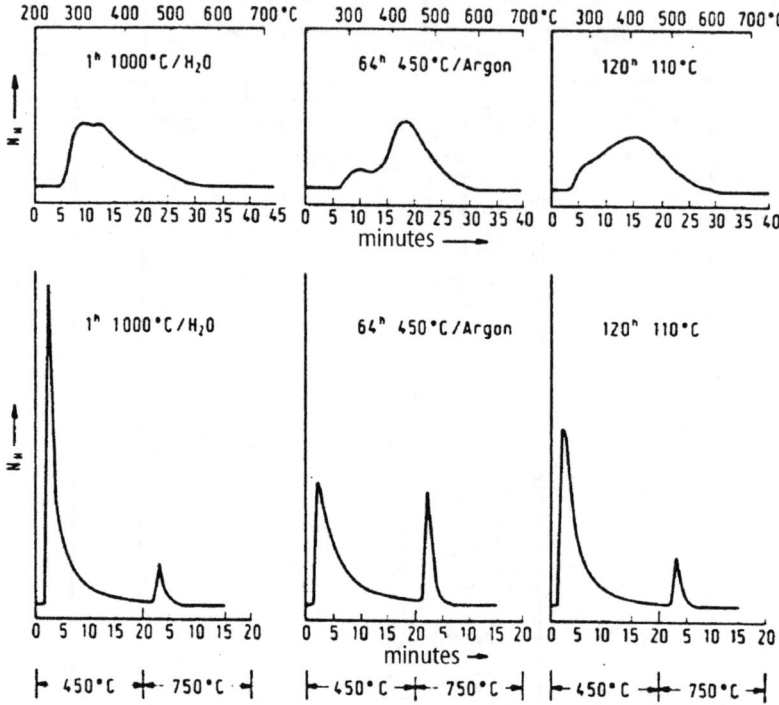

Fig. 6.23. Extraction curves of hydrogen reducible nitrogen (N_H) at 200–700 °C and 450/750 °C of a Fe-Mn-N alloy with 1 m/m % Mn and 50 mg/kg N after different thermal treatments

As a final example of the coming together of material development and material analytics, reference is made to investigations of "Galfan", an alloy galvanized sheet already mentioned above. In this case, examinations of the influence of composition and cooling rate on the microstructure and of the relationship between surface structure and primary crystal amount formed the basis for the safe production of this new surface coated material.

6.7.3
Surface Technology and Surface Analytics

6.7.3.1
Importance of Surface Analytical Methods

The first and direct impression of a product or of a material is effected by its surface. It causes in the observer a spontaneous, automatically subjective judgment of the properties of this product, for example with regard to the uniformity of the surface. The naked eye observations (dots, stripes, discolorations) are used for qualitative, not always objective, formation of an assessment. This impression is naturally not an adequate characterization if one considers that the mate-

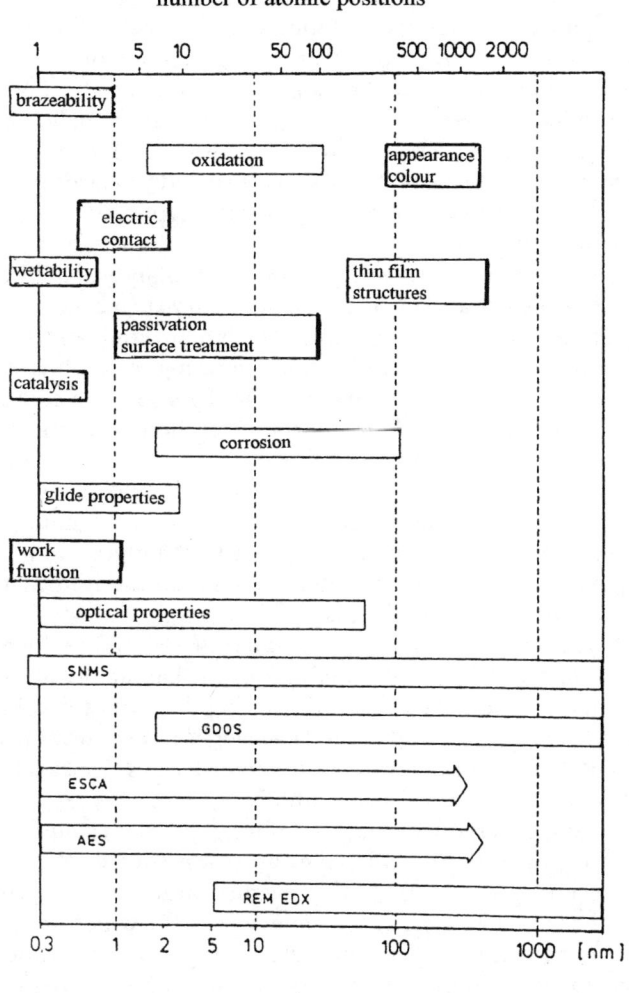

Fig. 6.24. Application ranges of different surface analytical methods

rial surface observed is subjected to further process steps and is therefore to be assessed regarding its useability for the desired technical process. Therefore, objective (verifiable) standards must be found in order to enable description of the surface of a product. Thus, the importance of surface analytics for material development and production becomes directly recognizable. The interest in this problem posed is naturally not restricted to the material producer; the processor and the consumer have the same wishes where the homogeneity of the properties and of the appearance of the product, in this case of the surface properties, is concerned.

Therefore for some years surface analytics has been finding its way into production-accompanying materials analytics, e. g. for steel and coated materials. The *scientific* interest is thereby aimed at the clarification of process sequences and reaction mechanisms with the *commercial* aim of the optimization of process and product as well as the establishing of causes in the case of complaints about quality. For protection from corrosion and, possibly, to improve processability, sheet and black plate are surface coated to a large extent and to an increasing degree – as already mentioned – by the application of metallic layers or organic coating agents.

The variety of coating materials requires various analytical methods for characterizing the coatings and layers near the boundary surfaces.

By combination of various analytical methods an overall analytical statement is possible which permits conclusions about operating conditions and operating sequences or unusual stresses on the surface in the event of a complaint. Moreover, surface analytics frequently supports the technical solution of chemical problems such as the optimization of galvanic baths.

At this point, the various electron spectroscopic and mass spectrometric methods of surface analytics and their fields of application cannot be dealt with in more detail. Here reference should be made to the literature [72–76] which, besides the physical basis, contains extensive data on the informativeness and the applicability on technical problems.

Glow discharge optical emission spectrometry (GDOES) [77] which enables, within a few minutes, multielement analysis for the characterization of surfaces of metallic materials and for the recording of depth profiles [78] has proved to be useable with very great success [79]. However, it has been shown that multi-method concepts are increasingly necessary in order to solve problems encountered in industrial practice. Of the methods available [74, 75], two processes which have proved successful in a large number of problem solutions in the field of steel processing are selected and looked at in more detail.

Optical emission spectrometry with glow discharge excitation (GDOES) and secondary neutral particle mass spectrometry (SNMS), which are characterized by good limits of detection, high sputter rates (sample abrasion) and a very good time response for the generation of depth profiles, have proved to be suitable investigation methods in industrial practice. By these methods relatively low excitation energies are applied to the sputter process (<500 eV) so that sputter-induced effects which bring about a change in the composition of the samples by the ion bombardment play a subordinate role. The practical depth resolution of GDOES is 5 nm, that for SNMS 0.5 nm. According to the coating processes applied today the surface questions to be handled, for example in the steel industry with priority, are in depth ranges of 1–500 nm (Fig. 6.24) such as in the case of oxidation and passivation layers, questions of surface treatment or corrosion phenomena. Additionally, there are metallic layers as, for example, in the case of hot galvanized material which can clearly exceed the 10 μm range.

SNMS in particular permits the determination of the composition of the thinnest layers with limits of detection into the μg/g range [80]. By recording of depth profiles, measurement of the concentration curve of one or several elements as a function of the distance from the surface, quantitative statements on element enrichments in areas close to the surface and in interface layers, the transition between the coating and the base material, can be made [81, 82].

6.7.3.2
Application of Glow Discharge Emission Spectrometry

The sample arranged as the cathode seals up the evacuated glow lamp (Fig. 6.25), which serves as a spectrochemical excitation source, from the atmosphere. After a suitable argon pressure of a few millibars has been reached a glow discharge is

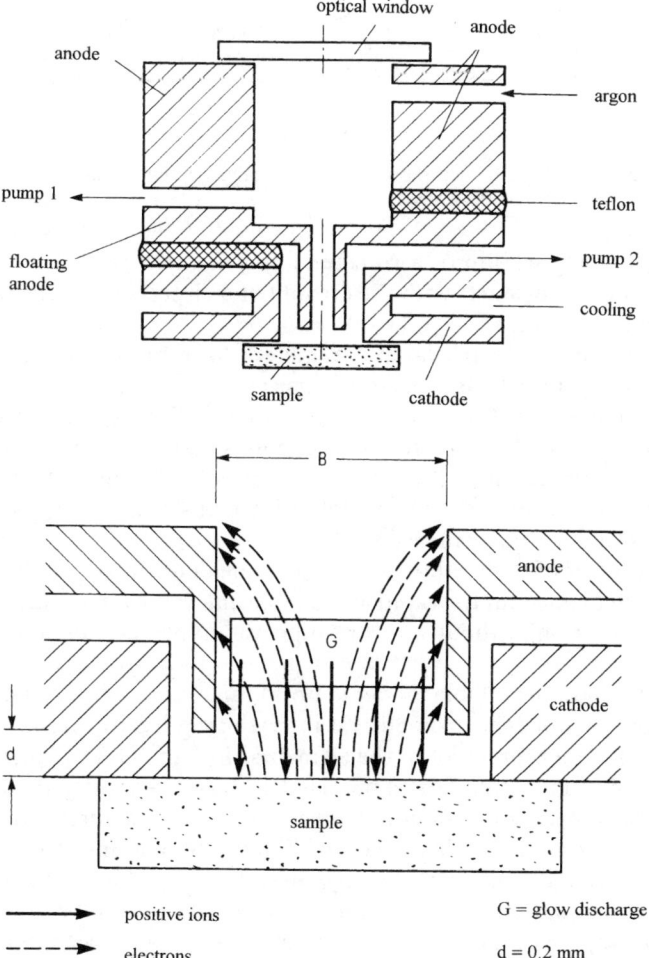

Fig. 6.25. Glow discharge lamp according to Grimm as spectroscopic excitation source (schematically)

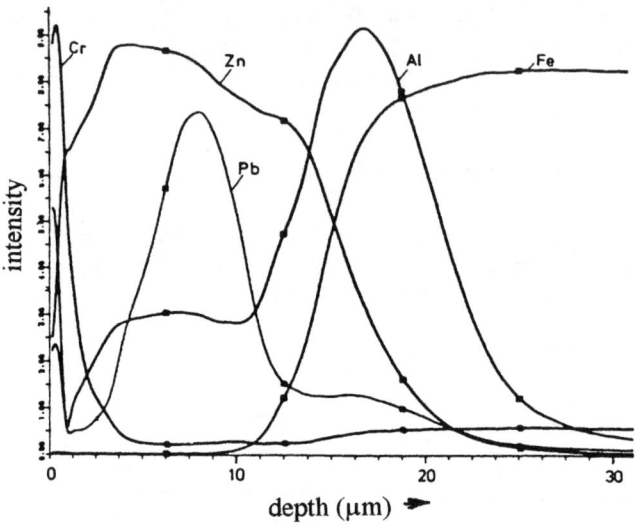

Fig. 6.26. Example of a GDOES depth profile: hot dip galvanized metal sheet (by permission of Fried. Krupp AG Hoesch-Krupp, Essen)

built up by the application of a voltage of, for example, 1000 V. The analytical cycle now proceeding starts with the erosion of a circular surface with a diameter of 8 mm by argon ions.

By selection of the electric lamp parameters, for example in the case of steel, erosion rates of 5 nm/s to approx. 100 nm/s can be achieved.

The eroded atoms of the sample diffuse into the discharge and are excited to radiate by impacts with electrons and argon ions and emit the ultraviolet radiation which is characteristic of them [83]. The glow light illuminates the entry slit of an optical spectrometer with the help of which several analysis lines of the analyte are simultaneously measured.

The advantages of the method are the possible detection of all the (technically relevant) elements and the rapid and simple analysis even of thick layers (μm range) and working without an ultra-high vacuum. The lateral resolution is <5 mm.

An analysis system suitable for depth profile analysis contains as its core component a high vacuum spectrometer for the wavelength range 110–450 nm which also enables the measurement of the elements H, N and O. The glow lamp serving as an excitation source can be operated with constant current and constant voltage, but also with constant power at various argon pressures. The measuring signals for, for example, 32 elements can be inquired of at intervals of 1 ms, presented and stored in real time. Thereby, correction possibilities for the measuring conditions or selection criteria for following steps can be read off immediately. In this way success is achieved in measuring surface layers deviating from the matrix composition and those interesting alterations in the area close to the surface which are due to the influence of process engineering factors, within a few minutes.

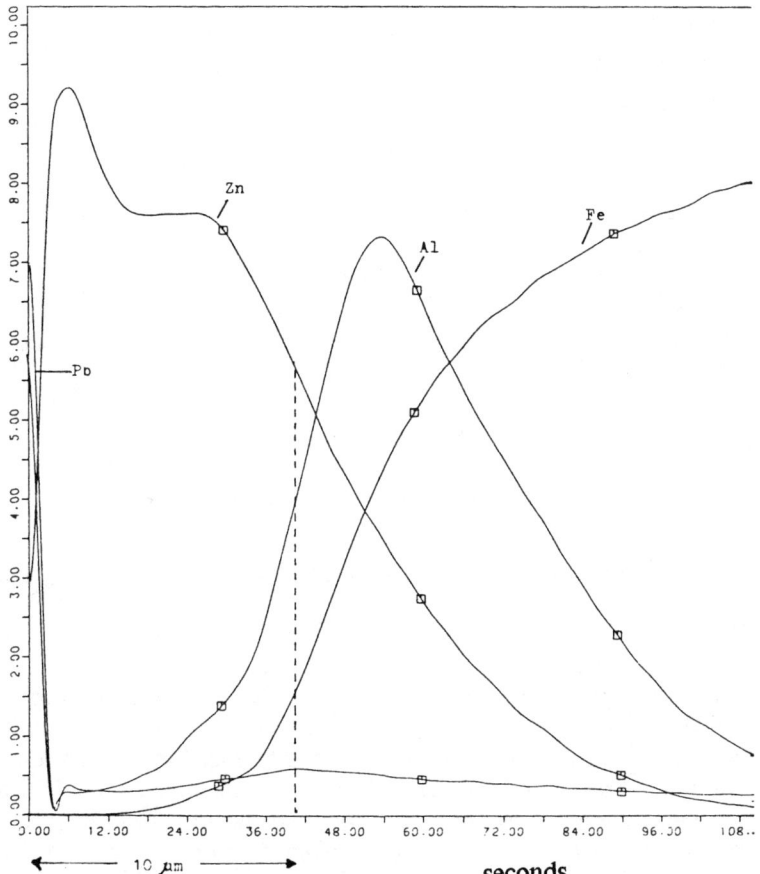

Fig. 6.26. (continued)

Such an analysis system permits, with its high time-resolved signal processing, the recording of time/intensity curves which are interpreted as depth profiles (Fig. 6.26). Enrichments are recognized as intensity maxima and impoverishments as intensity minima. By cathode atomization a crater is formed with an almost vertical crater edge. The transition to the bottom is marked by a small hollow. The crater bottom is macroscopically largely plane parallel to the original surface.

The information obtained in the form of time/intensity functions requires interpretation. The ratio of crater diameter (e.g. 8 mm) to the erosion depth per ms (=0.1 nm) as well as the macroscopic configuration of the crater (influenced by the analytical parameters) enable one to speak of geometrical layer sequences. The change in the density of the first atom layers does not, however, make possible an exact differentiation of these top layers. The qualitative assess-

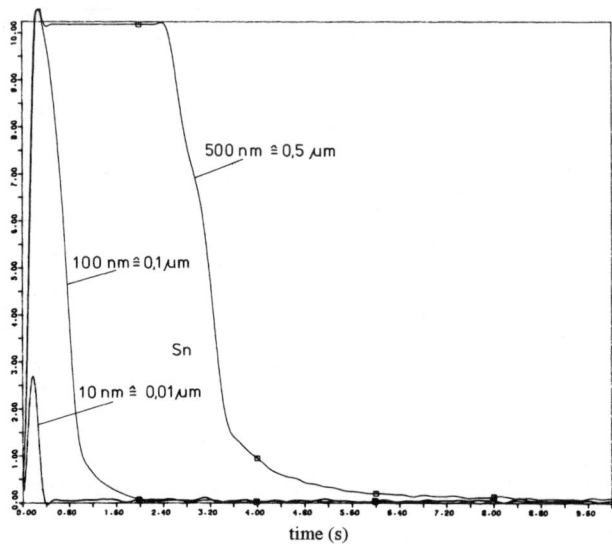

Fig. 6.27. Determination of tin layer thickness on steel (by permission of Fried. Krupp AG Hoesch-Krupp, Essen)

ment of GDOES depth profiles is comparatively simple taking into account the spectrometric parameters.

The quantifying of thin layers is essentially influenced by adsorption effects and the sputter characteristics of the surface composition. A prerequisite for the quantitative assessment of real samples is degreasing with, for example, toluene and storage of the samples at temperatures of 50 °C in order to avoid condensation. As the boundary surfaces of real samples contain water and adsorbed gases such as carbon dioxide and hydrocarbons, these substances must be removed before analysis as otherwise wrong depth profiles would result. For quantitative statements, for example on the layer thickness, the knowledge of the sputter characteristics is an essential prerequisite. For layer systems, such as galvanized sheets, scaling and protective layers, quantitative evaluations are possible [84–87]. For the assessment of such layers knowledge of the erosion rates of the various elements is necessary. That assumes the examination of appropriately coated samples, from which the coated layers are chemically removed and quantitatively determined by ICP spectrometry. Taking the density into account the layer thickness can be calculated (Fig. 6.27), while the erosion rates can be determined from the pattern of the GDOES curves.

The depth profiles thus obtained permit conclusions to be drawn regarding process sequences and possibly for influencing the surface properties of industrially produced products. In galvanic coating technology, continuous layer thickness measurement plays an important part for process control which can possibly be integrated into a computer-assisted quality assurance system [88]. Here, X-ray fluorescence spectrometric in-line methods are applied.

Examples of Application from Actual Practice of an Analytical Laboratory in the Steel Industry

With the help of the analysis system described, numerous questions relating to surface technology and materials science can be dealt with in short time ("fast product information"). On this, some examples are now given.

During the annealing of sheet steel, characteristic enrichments arise as a function of the annealing temperature and the protective gas composition. These alterations in the area close to the surface result from a reaction of the steel surface with constituents of the gas phase and by diffusion of elements from the matrix. At the same time, besides the oxidation of the iron, oxides of the Al, Si, Mn, Cr and P are formed as a function of the dew point of the gas. These enrichments can, for example, in the case of sheets for enamelling purposes, influence the pickling erosion and the *enamel adhesion*. Elements with an oxygen affinity such as Mn and Al are subject to, depending on the temperature, remarkable enrichments and impoverishments in the range 20–100 nm which have an influence on the following process steps. Therefore GDOES investigations are used for the optimization of the process.

Mixed lead coating (Ternal®) is one of the processes for the corrosion protection of steel surfaces. In connection with the investigation of the corrosion protection effect of the Pb/Sn layer applied the question of the distribution of the alloying elements in the coating should be answered. The GDOES depth profile analysis showed that the tin – as required for achieving reliable corrosion protection (avoidance of local element formation) – is distributed homogeneously within the lead.

Among the coating processes of steel sheets *hot dip galvanizing* has a special position. The corrosion protection thus intended is naturally only possible in the case of adequate adhesion of the zinc layer which is influenced by the formation of an Fe_2Al_5 layer at the steel/zinc interface. In order to achieve good zinc adhesion the galvanizing conditions must be defined in such a way that the Al, the amount of which in the zinc bath can, for example, be 0.2% m/m and more, diffuses into the coating on the steel/zinc phase boundary in order to form there a homogeneous Fe_2Al_5 layer over the whole area. This Fe_2Al_5 layer hampers the formation of an excessively thick Fe/Zn alloy layer and thus improves the adhesion. The GDOES was applied in this case to provide findings on the Al distribution. The depth profiles recorded provided decisive indications for the optimization of the process and the improvement of the product, for example by following annealing.

6.7.3.3
Investigations of Coated Materials with Secondary Neutral Particle Mass Spectrometry

As GDOES mainly allows the analysis only of metallic surfaces, for a wide industrial product range the demand for the investigation of, e.g. oxidic layers (corrosion problems; scaling formation), the determination of the chemical binding conditions in surface layers and of plastic-coated materials is soon made. Such tasks can generally only be handled and solved in a close combination of methods. This necessity to apply multimethod concepts is met by modern equipment development as a result of the creation of new items of combined equipment [89]. For these problems indicated, for example, an analyzer can be used which permits both SNMS (secondary neutral particle mass spectrometry) and SIMS (secondary ion mass spectrometry) investigations.

In the case of SNMS the erosion of the sample material is effected by bombardment with argon ions which are either produced in an HF plasma or are introduced by means of an external ion source (Fig. 6.28). These primary ions (energy 0.1–10 keV) produce an impact cascade in that area of the solid close to the surface, in the course of which atomic and molecular fragments are emitted from the surface. After the separation of the secondary ions the neutral particles are reionized by electron impact and detected with a mass spectrometer. The distribution of the fragments is characteristic of the chemical composition of the sample surface [90].

As advantages of SNMS the following can be mentioned:

- detection of all elements including hydrogen;
- high depth resolution by low energetic ion bombardment;
- direct relation between the intensities of the SNMS signals and the chemical composition of the sample;
- quantitative depth profiles;
- low matrix effects;
- possibilities of the investigation of thinnest samples and foils as well as of small, non-plane samples;
- examination of semiconductors and non-conductors.

The intensity of an SNMS signal depends on the primary ion current, the partial atomization yield, the reionization probability, a geometry factor and the secondary ion formation probability of an emitted species. The reionization probability is a specific constant of the apparatus in the case of reproducing conditions.

A quantitative evaluation finally becomes possible via a relationship in which the ratio of the concentrations of two different components in a sample corresponds to the ratio of the corresponding measured intensities and relative "sensitivity factors" (specific constant of the measuring equipment):

$$\frac{C_x}{C_{ref}} = \frac{I_x}{I_{ref}} \cdot \frac{D_{ref}}{D_x}$$

C_x and C_{ref} are the contents of the element to be determined and of the reference element, I_x and I_{ref} are the corresponding intensities and D_x and D_{ref} are the associated sensitivity factors. The ratio D_{ref}/D_x is usually termed a relative element sensitivity factor. By investigations of samples of known composition the sensitivity factors can be determined so that samples of unknown composition can be quantitatively analyzed [91].

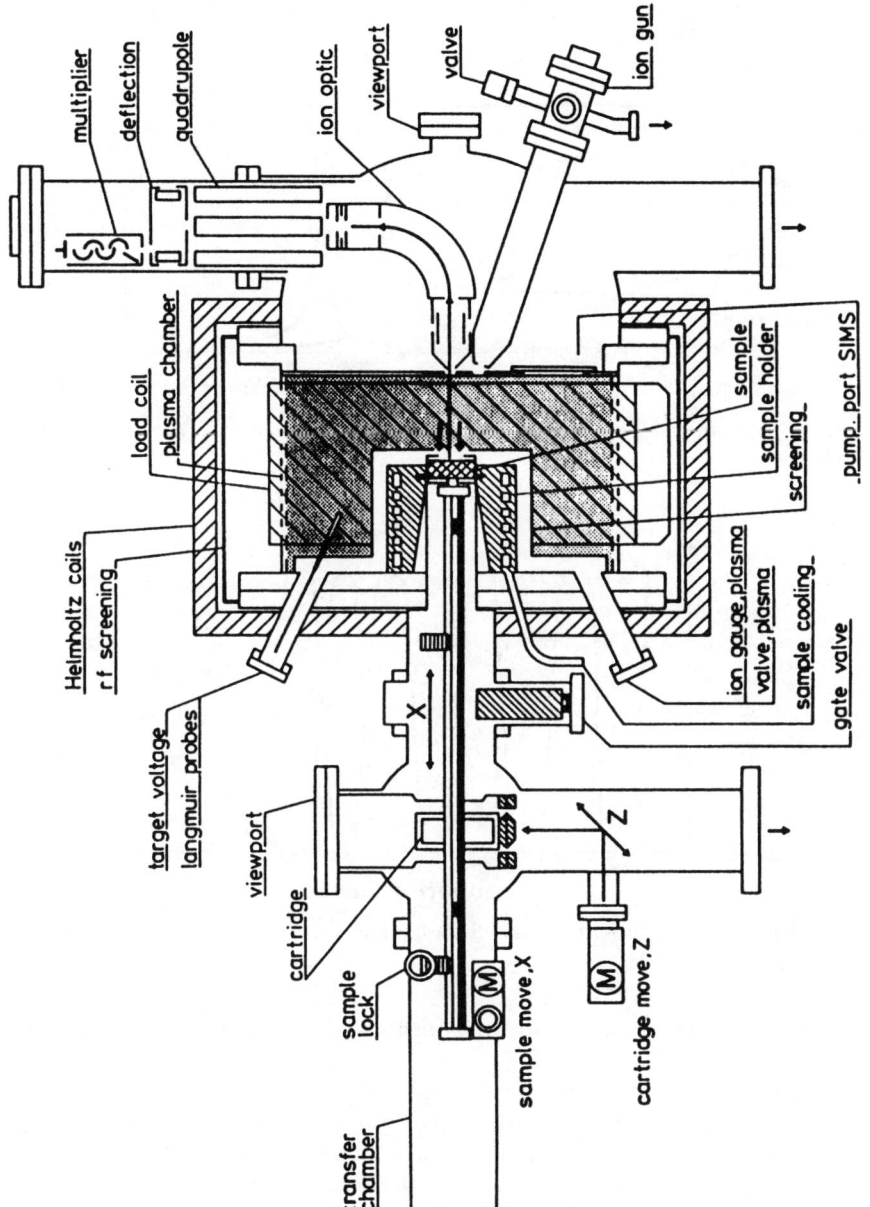

Fig. 6.28. Scheme of an SNMS analyzer (by permission of Fried. Krupp AG Hoesch-Krupp, Essen)

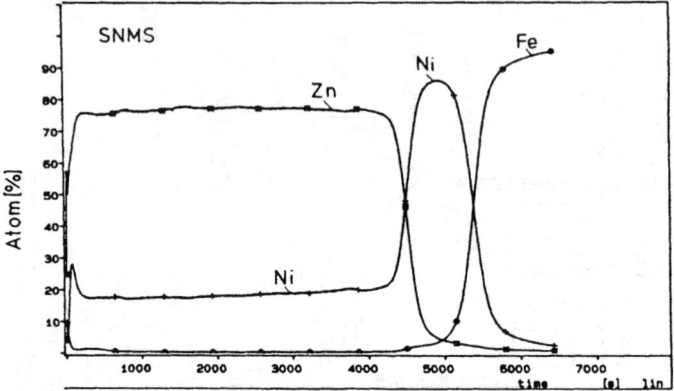

Fig. 6.29. SNMS depth profile of a multilayer system (by permission of Fried. Krupp AG Hoesch-Krupp, Essen)

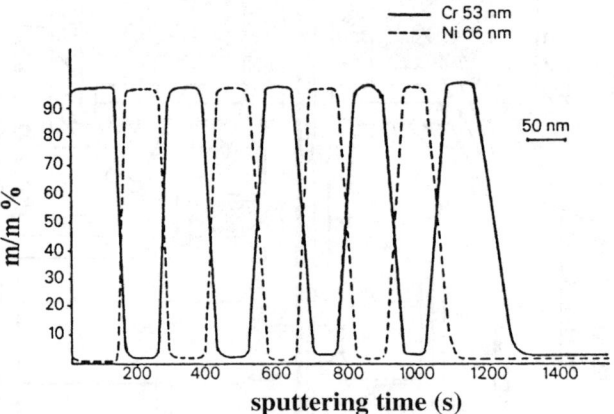

Fig. 6.30. SNMS depth profile of a titanium nitride layer on steel (by permission of Fried. Krupp AG Hoesch-Krupp, Essen)

For the characterization of industrial tasks some examples should be mentioned.

In the first case, besides a qualitative depth profile analysis the determination of the quantitative composition of a *Zn-Ni layer* coated on a steel material was required. On this material first of all a Ni layer had been deposited and then the Zn/Ni alloy layer was applied on top of it. GDOES first led to the qualitative findings on the layer structure, while SNMS (Fig. 6.29), on the one hand, confirmed the GDOES result (example of the assuring of findings by a second independent method!) and, on the other hand, provided the chemical composition of the alloy layer.

A further area is the investigation of *multilayer systems* by SNMS depth profile analytics. Due to the low sputter energy no mixings in the various layers occur at all (Fig. 6.30).

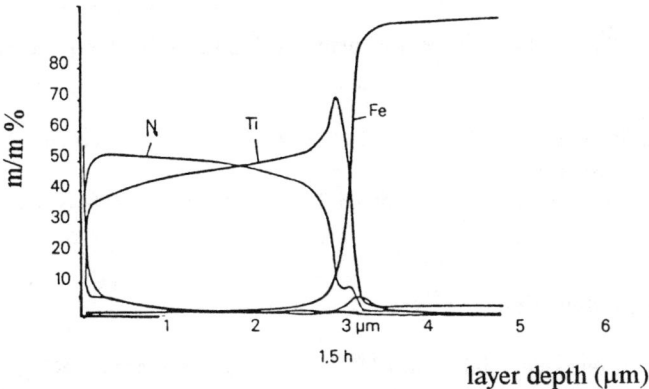

Fig. 6.31. SNMS depth profile of a titanium nitride layer on steel (by permission of Fried. Krupp AG Hoesch-Krupp, Essen)

In industrial practice *nitrided cases* are assessed by SNMS depth profile analysis (Fig. 6.31). Both the nitrided case thickness and the alterations of the quantitative composition within the layers can be determined and thus provide interesting metallurgical insights and material scientific information.

In conclusion it remains to be stated that the field of surface analytics is experiencing constant expansion. This includes, for example, the high-resolution depth distribution analysis of trace elements, the quantitative distribution analysis in thinnest multi-layer systems, the quantitative trace analysis in monolayers or three-dimensional stereometric analysis [92] as well as the surface microanalysis of solids [93].

6.8
Suppliers of Sampling, Sample Preparation, Analysis and Automation Systems

6.8.1
Sampling Probes

Heraeus Electro-Nite GmbH, Im Stift 6–8, D-58119 Hagen
Mincon-Sampler-Technik GmbH, Heinrich-Hertz-Strasse 30, D-40699 Erkrath
Soled Meßgeräte GmbH, Sympherstrasse101, D-47138 Duisburg

6.8.2
Pneumatic Tube Systems for Sample Transport

Maschinenfabrik Herzog GmbH & Co, Auf dem Gehren 1, D-49086 Osnabrück
Krupp Polysius AG, Graf Galen-Strasse 17, D-59269 Beckum
Pfaff AQS GmbH, Postfach 240228, D-42232 Wuppertal

6.8.3
Automated Sample Preparation by Cutting, Grinding or Crushing and Pressing Respectively

Maschinenfabrik Herzog GmbH & Co (see above)
Pfaff GmbH (see above)
Spectro Analytical Instruments GmbH, Boschstrasse10, D-47533 Kleve

6.8.4
Laboratory Automation

Fisons Instruments Vertriebs-GmbH, Peter-Sander-Strasse 43, D-55252 Mainz
Maschinenfabrik Herzog GmbH & Co (s.a.)
Hewlett-Packard GmbH, Hewlett-Packard-Strasse 8, D-76337 Waldbronn
Krupp Polysius AG (see above)
Pfaff GmbH (see above)
Polycon Gesellschaft für Laborautomation mbH, Bahnhofstrasse15, D-47559 Kranenburg
Spectro Analytical Instruments GmbH (see above)

6.8.5
Calibration Samples/Reference Materials

BAM – Bundesanstalt für Materialforschung und -prüfung, Unter den Eichen 87, D-12205 Berlin
BAS Bureau of Analytical Samples Ltd., Newham Hall, Newby, Middlesborough, Cleveland, England TS 8 9 EA, UK
Breitländer-Eichproben und Labormaterial GmbH, Postfach 8046, Hans-Sachs-Strasse 12, D-59077 Hamm
H.R.T. Labortechnik GmbH, Klausnerring 17, D-85551 Kirchheim
U.S. Department of Commerce, National Bureau of Standards (NBS), Office of Standard Reference Materials, Rm. B 311 Chemistry Bldg., Gaithersburg, MD 20899, USA

6.8.6
Optical Emission Spectrometer

Baird Europe B.V., Produktieweg 30, NL – 2382 PC Zoeterwoude
Belec GmbH, Hamburger Strasse 12, D-49124 Georgsmarienhütte
Fisons Instruments Vertriebs-GmbH (see above)
OBLF GmbH für Elektrotechnik und Feinwerktechnik, Salinger Feld 44, D-58454 Witten
Shimadzu Europa GmbH, Postfach 290260, D-47262 Duisburg
Spectro Analytical Instruments GmbH (see above)
Carl Zeiss Jena GmbH, Tatzendpromenade 1 a, D-07745 Jena

6.8.7
X-Ray Fluorescence Spectrometer

Atomika Instruments GmbH, Brukmannring 6, D-85764 Oberschleißheim
Fisons Instruments Vertriebs-GmbH (see above)
Philips Industrial Electronics Deutschland GmbH, Miramstrasse87, D-34123 Kassel
Oxford Instruments Deutschland GmbH, Kreuzberger Ring 38, D-65205 Wiesbaden
Siemens AG, Röntgenanalytik, Postfach, D-76181 Karlsruhe
UNICAM Analytische Systeme GmbH, Korbacher Strasse 75–77, D-34132 Kassel

6.8.8
Analyzers for the Determination of Carbon, Sulphur, Nitrogen, Oxygen and Hydrogen in Metals

Fisher-Rosemount GmbH & Co, Wilhelm-Rohn-Strasse 51, D-63450 Hanau
LECO Instrumente GmbH, Benzstrasse5 b, D-85551 Kirchheim
Ströhlein GmbH & Co, Girmeskreuzstrasse55, D-41564 Kaarst

6.8.9
Container/Cabin Laboratories

Baird Europe B.V. (see above)
Fisons Instruments Vertriebs-GmbH (see above)
Maschinenfabrik Herzog GmbH & Co (see above)
Spectro Analytical Instruments GmbH (see above)

6.8.10
Mobile Spectrometer

Baird Europe B.V. (see above)
Belec GmbH (see above)
Fisons Instruments Vertriebs-GmbH (see above)
Metorex GmbH, Königsteiner Str.98, D-65812 Bad Soden/Ts.
Leeman-Labs Analysegeräte GmbH, Postfach 1326, D-57258 Freudenberg
Outokumpu KM-Analytik GmbH, Königsteiner Strasse 98, D-65812 Bad Soden
Oxford Instruments GmbH, Postfach 4509, D-65035 Wiesbaden
Polycon Gesellschaft für Laborautomation mbH (see above)
Spectro Analytical Instruments GmbH (see above)

6.8.11
Remarks

The preceding compilation cannot make any claim to completeness. For further information reference should be made to the literature and the market surveys published in "Nachrichten für Chemie, Technik und Laboratorium", the news of the Gesellschaft Deutscher Chemiker (GDCh) (Society of German Chemists). Further information can be found in the References to Chaps. 3 and 4.

References

1. Johannsen O (1953) Geschichte des Eisens, 3. Aufl., Verlag Stahleisen, Düsseldorf
2. Koch KH (1987) Mikrochim. Acta (Wien) I:151
3. Wiegand H (1977) Eisenwerkstoffe- Metallkundliche und technologische Grundlagen, Physik, Verlag Chemie, Taschentext 55, Weinheim
4. (1992) Stahlschlüssel, 16. Aufl., Verlag Stahleisen, Düsseldorf
5. Kemp N (1968) Fresenius' Z Anal Chem 240:303
6. Flock J, Koch KH, Ohls K (1991) Stahl u. Eisen 111(8):103
7. Ono A, Chiba K, Saeki M, Ninbe H, Kasai S (1985) Trans ISIJ 25:B39
8. Nomomura E, Kotani N, Tokuda T,Yoshida Y, Yabata T, Narita K (1985) Trans. ISIJ 25:B37
9. Tsunoyama K, Tanimoto W, Hisada H, Asakawa H (1985) Trans. ISIJ 25:B41
10. Ichihara K, Janke D, Engell H-J (1986) Steel Research 57:166
11. Küppers A (1986) Steel Research 57:295
12. Koch KH (1985) Stahl u. Eisen 105:1176
13. Koch KH (1984) Spectrochim. Acta 39B:1067
14. Koch KH, Wünsch H (1982) Stahl u. Eisen 102:497
15. Bardenheuer F, Ansmann W, Pfeiffer A (1982) Stahl u. Eisen 102:917
16. Fiege L, Kaiser H-P, Delhey H-M, Schäfer H (1985) Stahl u. Eisen 105:1443
17. Baker R (1985) Proc 38th BSC/BISPA-Chemists' Conference, BSC, Research and Development Department, Scarborough, p 9
18. Petesch P, Böhm G (1993) Stahl u. Eisen 113(10):77
19. Fujino N, Matsumoto Y, Yoshihara M, Tarui M (1982) Tetsu-to-Hagane 68:2585
20. Pluschkell W (1979) Stahl u. Eisen 99:398
21. Janke D, Hagen K, Hammerschmid P, Dittert D, Lindenberg H-U, Weber L (1987) Stahl u. Eisen 107:537
22. Koch KH, Ohls K, Flock J (1989) Proc 42nd BCS/BISPA-Chemists' Conf. Scarborough, p 79
23. Sugihara T, Saiton K, Gouda A, Koishi S, Hata T (1988) Kawasaki Steel Techn Rep 19:87
24. Thierig D, Unger H, Dehrendorf H, Theiß H-J (1984) Fresenius' Z Anal Chem 319:10
25. Schmitz L, Loose W, Koch KH (1975) Fresenius' Z Anal Chem, 276:111
26. DIN 51 418, Part 1 Draft (1993): X-ray spectrometry; X-ray emission and X-ray fluorescence analysis (XRF); definitions and basic principles. DIN 51 418, Part 2: X-ray spectrometry; X-ray emission and X-ray fluorescence analysis (XRF); definitions and basic principles for measurement, calibration and evaluation of results
27. Ehrhardt H (1981) Röntgenfluoreszenzanalyse. Anwendungen in Betriebslaboratorien. VEB Deutscher Verlag f. Grundstoffind., Leipzig
28. Lajunen LHJ (1992) Spectrochemical analysis by atomic absorption and emission. The Royal Society of Chemistry, Blackhorse Road, Letchworth, Herts SG6 1HN, UK
29. Broekaert JAC (1994) In: Ullmann's encyclopedia of industrial chemistry, vol B 5. VCH Verlagsges., Weinheim, p 559
30. Habel K, Schlothmann B-J, Staats G, Thiemann E, Thierig D (1990) Stahl u. Eisen 110(5):53
31. Koch KH (1981) Arch Eisenhüttenwes 52:479
32. Koch KH (1981) In: G. Kraft (ed) Analysis of non-metals in metals. W de Gruyter, Berlin, p 3
33. Flock J, Koch KH, Ohls K (1989) Stahl u. Eisen 109:1223
34. Kollin G, Ziemens F (1972) Neue Hütte 17:556
35. Kipsch D (1972) Neue Hütte 17:111
36. Pfeiffer A, Reski H D, Thiemann E (1991) Stahl u. Eisen 111(5):103
37. Scheufler R, Brauner A (1989) Steel technology int. Sterling, London, p 361
38. Slickers KA, Schmitten G, Beer E, Klein N, Veit I (1988) Giesserei 75:H. 15/16:469
39. Summerhill BD, Murdoch TD (1992) In: Nauche R (ed) Progress of analytical chemistry in the iron and steel industry. Commission of the European Communities, EUR 14 113 EN, Luxemburg, p 333
40. Slickers KA (1993) Stahl u. Eisen 113(7):103

41. Dimitrov S, Ranganathan S, Weyl A, Janke D (1993) Steel Research 64(1):63
42. Whiteside IRC, Jowitt R (1992) In: Nauche R (ed) Progress of analytical chemistry in the iron and steel industry. Commission of the European Communities, EUR 14 113 EN, Luxemburg, p 135
43. Carlhoff C, Kirchhoff S (1992) In: Nauche R (ed) Progress of analytical chemistry in the iron and steel industry. Commission of the European Communities, EUR 14 113 EN, Luxemburg, p 150
44 Koch KH, Kopineck H-J, Meininghaus F, Tappe W (1986) DE-Patent 37 68 345
45. Kudermann G, Lührs C (1989) Erzmetall 42:163
46. Kopineck H-J (1993) Stahl u. Eisen 113(11):73
47. Koerfer E, Berstermann W, Schäfke R (1986) Stahl u. Eisen 106:167
48. Koch KH, Flock J (1991) Hoesch-Ber. Forsch. Entw. H. 1:19
49. Berstermann W (1992) Stahl u. Eisen 112(8):95
50. Kopineck H-J, Tappe W (1989) Stahl u. Eisen 109:1151
51. Kopineck H-J, Tappe W (1970) Mikrochim. Acta Wien Suppl. IV:48
52. Burek J, Gantner E, Ache H J (1993) Fresenius' J Anal Chem 346:671
53. Paatsch W (1990) Galvanotechn. 81:3852
54. Abe T, Yasui N, Yamato K, Takatoku Y, Toumori T, Kurozumi S (1985) Transactions ISIJ 25:B49
55. Kondo K, Shibazaki T, Iwanuma K, Kimura T, Mashino Y, Saizu M, Sekigushi H (1985) Transactions ISIJ 25:B47
56. Broekaert J A C, Tölg G (1987) Fresenius' Z Anal Chem 326:495
57. Jenkins R (1994) In: Ullmann's encyclopedia of industrial chemistry, vol B 5. VCH Verlagsges., Weinheim, p 675
58. Kellner R, Mermet J-M, Otto M, Widmer M (1998) Analytical chemistry. Wiley VCH, Weinheim
59. Broekaert JAC (1990) Analytiker-Taschenbuch, Bd. 9, Springer, Berlin Heidelberg New York, p 127
60. Grasserbauer M, Wer H W(1991) Analysis of microelectronic materials and devices. Wiley, Chichester
61. Stingeder G (1992) Fresenius' J Anal Chem 343:771
62. Ortner HM (1992) Fresenius' J Anal Chem 342:695
63. Blödorn W, Lück J (1992) Fresenius' J Anal Chem 343:705
64. Kudermann G, Blaufuß KH, Lührs C, Vielhaber W, Collisi U (1992) Fresenius' J Anal Chem 343:734
65. Lange, Blödorn W (1981) Das Elektronenmikroskop – TEM und REM. G Thieme, Stuttgart
66. Grabke HJ, Erhart H, Möller R (1983) Mikrochim. Acta 10:119
67. Pohl M, Oppolzer H, Schild S (1983) Mikrochim Acta Suppl 10:281
68. Grasserbauer M (1985) Fresenius' J Anal Chem 322:105
69. Swinburn DG (1974) The separation and determination of nitride phases present in steel. BSC Open, CDL/CAC/48/74
70. Flock J, Koch KH, Ohls K (1988) Steel Research 59:1
71. Kretschmer M, Koch KH (1977) Radex-Rundsch, p 301
72. Hofmann S (1980) Analytiker-Taschenbuch, Bd. 1, Springer, Berlin Heidelberg New York, p 287
73. Grasserbauer M, DudeKHJ, Ebel MF (1986) Angewandte Oberflächenanalyse. Springer, Berlin Heidelberg New York
74. Hantsche H (1993) Jahrb. Oberflächentechn., Bd. 49, Metall, Berlin Heidelberg, p 381
75. Rivière JC (1990) Surface analytical techniques. Clarendon, Oxford
76. Briggs D, Seals M (1990/1992) Practical surface analysis", 2. Aufl., 2 Bde. Wiley, Chichester
77. Berneron R, Charbonnier JC (1981) Surf Interface Anal 3:134
78. Kretschmer M, Koch KH, Grunenberg D (1983) Fresenius' Z Anal Chem 314:226
79. Koch KH, Sommer D, Grunenberg D (1984) Radex-Rdsch. H. 3/4:437
80. Müller KH, Oechsner H (1983) Mikrochim Acta, Suppl 10:51
81. Oechsner H (ed) (1984) Thin film and depth profile analysis. Springer, Berlin Heidelberg New York, p 63
82. Leis F (1995) Nachr Chem Tech Lab 43:967
83. Laqua K (1991) Analytiker Taschenbuch, Bd. 10. Springer, Berlin Heidelberg New York, p 297
84. Bengtson A (1985) Spektrochim Acta 40 B:631
85. Rose E, Mayr P (1989) Mikrochim Acta I:197

86. Angeli J, Kaltenbrunner T, Androsch FM (1991) Fresenius' Z Anal Chem 341:140
87. Dessenne O, Quentmeier A, Bubert H (1993) Fresenius' J Anal Chem 340:345
88. Staib W (1989) Metalloberfläche 43:467
89. Berresheim K (1984) Fresenius' Z Anal Chem 318:661
90. Oechsner H, Stumpe E (1977) Appl Phys 14:43
91. Wucher A, Oechsner H (1989) Fresenius' Z Anal Chem 33:470
92. Grasserbauer M, Friedbacher G, Hutter H, Stingeder G (1993) Fresenius' J Anal Chem 346:594
93. Wucher A (1993) Fresenius' J Anal Chem 346:3

7 Process Analytics in the Semiconductor Industry

7.1
Introduction

7.1.1 Microelectronic Technology

The progress in microelectronics, in particular in silicon technology, still proceeds rapidly, since advancing miniaturization of modules enables a continuous cost reduction in terms of cost per bit of semiconductor memories [1]. This development is joined, however, with a drastically increasing process complexity. Ultra-pure silicon wafers represent the base material of microchip manufacture, where in the case of the 4 M DRAM generation (M DRAM=megabit dynamic random access memory) up to 400 process steps are required in order to produce a microelectronic component with integrated circuits [2]. Besides the large number of metallic, semiconducting and insulating thin films that are necessary for the operation of the device, an even larger number of auxiliary layers are used that must be removed after having fulfilled their purpose. In order to achieve production yields at economic levels, a high level of process maturity with high reliability and stability is an indispensable prerequisite. This claim can only be realized by completely characterized base materials and a tighter and comprehensive process control. Therefore, problem-oriented analytical methods have to be provided and applied from the beginning of the development of a completely new generation of semiconductor devices as well as in their mass production later on.

Already ultratraces of contamination such as residues of common construction materials of fabrication tools and the environmental pollution in and on the wafer can reduce the production yield decisively or render the product worthless after an expensive fabrication cycle of several months. Therefore, a continuous defect reduction is of the highest interest in order to gain cycle time and thereby to increase the capacity of the installations without any additional investment as well as to reduce the extent and the costs of expensive testing procedures. The necessary investigations must include, besides the starting materials, all required consumables and auxiliary materials [3], whereby the analytics represents among other things the primary tool of continuous improvement of processes and products [4]. International guidelines [5] and expert systems [6] regulate its goal-oriented application.

In the connection described it must be considered, however, that the analytically measurable parameter can usually only be regarded as an indicator for a process or product state, but not yet as the requested quality characteristic itself. Moreover, it is to be emphasized that, as in all fields of analytics and particularly in the ultra-trace range, only regular introspective cross-checking of the results by means of different analytical methods [7] and by frequent interlaboratory comparison studies (round robin tests) [8] can assure reasonable accuracy.

Here too, process analysts are expected to provide meaningful results in real-time and with a high degree of reliability (see also Sects. 1.1, 5.1 and 6.3.1). These extreme demands on analytical process monitoring and research of causes can only be fulfilled by the application of physical and physicochemical instrumental methods. In this regard classical analytical chemistry is restricted to sample preparation, dissolution, separation and pre-concentration as well as to the control of sample handling and environmental influences.

7.1.2
Semiconductor Technology

The manufacturing of semiconductor silicon begins with the extraction and chemical purification of the raw material quartz sand. The pure silicon oxide produced is afterwards converted into silane or chlorosilane. After distillative separation from the electrically active impurities, it follows the thermal decomposition of this gas at 1100°C. Polycrystalline silicon results that can be converted by different procedures into monocrystalline silicon. Despite the high crystalline order obtainable, disorders in the crystal lattice and impurities can occur [9].

The monocrystalline silicon rods obtained are sliced by means of diamond-coated saws ("wafering"), chemically etched, chemomechanically polished, and chemically cleaned. The purpose of this process sequence is the removal of mechanically damaged and contaminated surface layers and the increase of the required strength, flatness and cleanliness of the Si wafer.

Semiconductor silicon challenges the analyst in many ways. Metallic contaminations, mechanical microdefects and a combination of their interactions can considerably reduce the chip yield or render the processed wafer useless [10]. Therefore, a reasonable production of semiconductor silicon needs a closed-loop monitoring system which can contribute to a high degree to the production and quality assurance as well as to the continuous improvement of the whole process event. Most of the efforts pursue the metallic cleanliness of the starting materials, the process media, the environmental conditions and at least the final product. To be successful in diagnostics and problem solving, the analyst must possess – as also known from other fields of process analytics – a basic theoretical and a good practical knowledge of the charged materials and the technical processes involved [11].

7.1.3
Solar Technology

The direct conversion of sunlight into electric current by means of solar cells (photovoltaics) is assigned an important role within the utilization of renewable forms of energy. Solar cells based on silicon already possess today a high technical standard and can be used economically in particular cases for current generation. The realization that the fossil sources of energy are only available for a limited time and that therefore renewable forms of energy must be developed is nowadays widespread.

Photovoltaics, i.e. the direct conversion of light into electric current by means of semiconductor materials, has gained a firm place in the market during the last few years [12]. It was reported by the Fraunhofer-Institut für Solare Energiesysteme (ISE), Freiburg (Germany) that in 1996 photovoltaic modules for more than 80 MW of electric power output have been sold all over the world. The generation of current by solar cells is still uneconomic in those cases where, for example, households are connected directly to the public power supply network. Special opportunities for the photovoltaics exist in regions with lasting and intensive sun solarisation. Here, it provides the opportunity to bring electricity into remote areas of developing countries. The actual fields of application are the house and village current supply as well as communication technology (e. g. generation of current in distant telephone and telegraph or signal facilities).

Only solar cells based on mono- or polycrystalline silicon are adequate at present to meet the requirements regarding efficiency and long-term stability [12]. The ultra-pure silicon required as starting material is produced in a fluid bed reactor either by decomposition of silane or by reduction of trichlorosilane as intermediate product (see Sect. 7.1.2). The polycrystalline material obtained is converted afterwards into monocrystals – as already described above – by different procedures (e.g., zone refining or cost saving casting procedures). The Si rods obtained are sliced by sawing into thin discs, which are processed after cleaning and quality inspection to solar cells. Regarding mass production, more extensive research and development work is urgently needed, which can be performed intensively at universities and industrial research institutions thanks to public financial support.

Process monitoring and quality inspection in solar technology is less expensive than in microelectronics. Polycrystalline silicon consists of crystallites of different orientations (see also Sects. 6.5 and 6.7.1). Grain boundaries and crystal defects can impair the efficiency of solar cells made of polycrystalline silicon. Therefore, it is essential to make imperfections of crystals visible as early as possible. An effective quality inspection starts directly *after* the manufacturing of the discs and *before* the costly process sequence of the solar cell fabrication and follows the guidelines of a quality management conscious of costs (see also Sect. 9.1).

By local illumination using laser diodes and measurement of the decay period of photo-generated charge carriers the quality of the silicon discs prepared for the fabrication by sawing can be measured [12]. For the characterization of fabricated solar cells on the basis of polycrystalline silicon, topographical measuring methods may be used which take into account the lateral inhomogeneity of the material. The most important parameters that determine the efficiency of solar cells are short-circuit current and open-circuit voltage of the cell. By selective illumination of very small regions of a solar cell with laser light, short-circuit current topographies, for example, are obtained with a dispersion in the micrometer range. Topographical descriptions of the most important characteristics of solar cells of polycrystalline silicon allow one to recognize the areas of reduced efficiency and thereby offer the basis for aiming at product improvement.

7.2
Investigation of Silicon Substrates

It has been shown that the tasks of analytics in the semiconductor industry concern the following fields: characterization of silicon substrates, analytics of thin layers, analytical control of ultra-pure chemicals and chemical processes as well as the contamination control of processes and production plants. Inadvertent contamination by metallic impurities in device processing is recognized as a major yield limiting factor and reliability risk since the early days of silicon device technology. In particular transition metals of the 3d group, such as Fe, Ni, Co and Cu, play an extremely detrimental role in this respect [13]. Defects and faults caused by these chemical elements when manufacturing microelectronic circuits, e.g. because of diffusion or forming of silicides, as well as the variety of possible defects caused by other influences such as ion implantation and etching procedures, and their avoidance and reaction mechanisms cannot be dealt with in detail here; instead one must be referred to the detailed literature [10, 14, 15]. In this context, it should be generally stated that scanning electron microscopy (SEM) and transmission electron microscopy (TEM) (see also Sect. 6.7.2) have proven to be indispensable investigation methods for the identification and detection of these defects [16].

The detrimental impact of metal contamination on device manufacturing has two consequences: on the one hand, the metal impurities form compounds (precipitates) which then act as nucleation sites in the generation of extended defects in the silicon substrate [17]; on the other hand, metal impurities decorate eagerly existing extended crystal defects enhancing the electrical activity of these defects by orders of magnitude. A number of sources of metal contamination have been identified in device production lines [13]:

- furnaces and epitaxy reactors;
- ion implanters, plasma and reactive ion etching equipment;
- wafer handling equipment and tools;
- wet chemicals for cleaning and etching.

Contamination control at first consists of prevention of contamination, i.e. by application of appropriate, repeated wafer cleaning and by using improved process equipment and wafer handling tools, and adequate quality levels of water, gases and chemicals. However, all these precautions cannot exclude accidental metal contamination during device fabrication. Here, gettering techniques can help to remove or keep harmful metal contamination away from active device regions [18].

For monitoring purposes the analytical methods have to be rather simple, fast, preferably non-destructive and preparation-free, and suitable for automated operation in a cleanroom environment [19]. With respect to the different sources of metal contamination, surface analytical *and* bulk analytical techniques are necessary. The main features of important analytical methods are summarized in Table 7.1. In this case and in the following, the description of imaging methods [20] was deliberately abstained from, these being somewhat important for the fast characterization of silicon substrates as well as for the bulk and surface analysis.

The surface analysis is needed for monitoring of all process steps, where the danger of contamination on the two wafer surfaces exists. Therefore, it should be dealt with specifically here. Recently total reflection X-ray fluorescence analysis (TXRF) has become generally accepted in monitoring surface contamination [21] and was introduced into the daily routine [22]. This method is based on the total reflection of X-rays at planar, polished wafer surfaces, where the characteristic fluorescence radiation of the contaminating metals (and those of the Si matrix) is excited up to a penetration (information) depth of approximately 3 nm [23]. This method makes possible the automated analysis of series with computer-aided data recording and data reduction. The calibration is performed using a wafer contaminated with a known amount of Ni atoms. For this purpose, from a certified standard solution used for example for calibration of atomic absorption spectrometry (AAS), an amount corresponding to 1 ng Ni is pipetted on the centre of the wafer in a small droplet and dried.

In daily routine, commercially available TXRF systems are able to quantify contamination down to 10^{10} atoms/cm^2. In order to quantify economically in the lower ranges, the contamination elements have to be concentrated on the silicon

Table 7.1. Selected analysis techniques for metal contamination [10, 20]

Method	Detection limit (atoms/cm^2)	Analyzed surface	Sample through-put (wafer/h)
TXRF[a]	10^{10}–10^{11}	1 cm^2	1
VPD[b]/TXRF	10^8	wafer	4
VPD/GF-AAS[c]	10^9–10^{10}	wafer	4
VPD/ICP-MS[d]	10^8	wafer	4
ETV-ICP[e]	10^7	wafer	–[g]
NAA[f]	10^9–10^{13}	wafer	–[g]

[a] Total reflection X-ray fluorescence analysis
[b] Vapour phase decomposition
[c] Graphite furnace atomic absorption spectrometry
[d] Inductively coupled plasma / mass spectrometry
[e] Electrothermal vaporization / inductively coupled plasma spectrometry
[f] Neutron activation analysis
[g] Unknown

surface. For this purpose, Vapor Phase Decomposition (VPD) has been developed, i.e. the reaction of the native oxide layer on the Si wafer with fluoric acid performed in a closed VPD-reactor in a clean room environment [24]. The reaction product on the surface is then collected by scanning the surface with an ultra-pure droplet of 100 µl. For a consecutive TXRF measurement the droplet must be dried by ultra-pure nitrogen at one spot on the Si surface. This combination of TXRF with VPD is able to detect metals down to 10^7 atoms/cm^2 on 200 mm wafers. For measurement by means of Graphite Furnace Atomic Absorption Spectrometry (GF-AAS) the sample obtained by the VPD technique is diluted to a defined stock volume and transferred into the sample cup of the measuring device. Combining VPD with Inductively Coupled Plasma Mass Spectrometry (ICP-MS) for the analysis of the VPD residue reveals the possibility of increasing the delectability by 2–3 orders of magnitude. Another method in spectroscopic surface analysis is the combination of Electrothermal Vaporisation (ETV) with ICP spectrometry with limits of detection at a level of 10^7 atoms/cm^2 (Fe) [25]. The costly and time-consuming Neutron Activation Analysis (NAA) is no longer applied to the investigation of contaminations and is today used as an independent reference and calibration method.

7.3
Surface and Thin Layer Analytics

As shown in the preceding description, surface analysis already represents for the investigation of silicon substrates an indispensable tool for process technology. In the following, it will be demonstrated that surface and thin layer analytical methods play a key role among the various analytical procedures used today in semiconductor technology [26]. They are applied to the characterization and monitoring of the properties and the quality of the manifold thin films, such as semiconductor surfaces of microelectronic components, interfaces, and thin doping layers. In particular, surface characterization and control also constitutes a main task of this analytical branch, because the nature and quality of the surface onto which a thin film is deposited influences crucially very often the properties and quality of the film and in, the case of semiconductor technology, eventually the reliability of the devices. Therefore considerable attention is paid to surface characterization *prior to* and *after* processes such as film deposition, etching and cleaning. The same applies to the interfaces formed or occurring by stacking films upon one another.

For solving the manifold surface analytical tasks the following methods are applied today:

- Auger electron spectroscopy (AES) [27];
- X-ray photoelectron spectroscopy (XPS) [27];
- secondary ion mass spectrometry (SIMS) [27];
- plasma chromatography mass spectrometry (PCMS) [28];
- total reflection X-ray fluorescence analysis (TXRF) [27]; and in special cases
- Rutherford backscattering spectroscopy (RBS) [27].

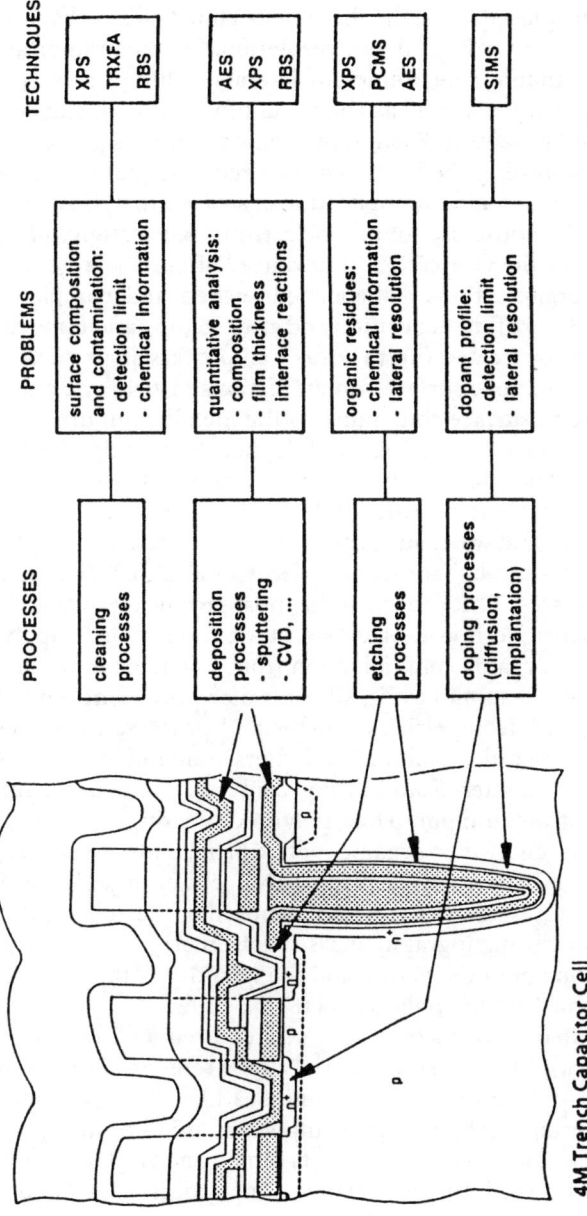

Fig. 7.1. Application of surface analysis in typical processes of microelectronics (by permission of Springer Verlag GmbH & Co KG, Berlin-Heidelberg)

Emphasis is to be placed upon the three first-named classical surface analytical methods which share a high depth resolution. Further important aspects of material analysis in microelectronics are lateral resolution, sensitivity for impurity and dopant analysis as well as chemical information due to complex chemical reactions and processes. Each of the named methods possesses its particular advantages regarding the indicated requirements [26]. Figure 7.1 illustrates the variety of problems and the application of surface analytical methods in connection with typical processes of microelectronic fabrication (4 M DRAM).

In the course of the fabrication of advanced circuits a large number of thin inorganic and organic films is deposited, etched and completely or locally removed again. Before film deposition and after etching and removal, the surface has to be cleaned or treated (up to 30% of all process steps consist of cleaning steps) in order to remove surface contaminations or to create a desired state of the surface. Among surface contaminants the metallic impurities, in particular transition metals such as Fe, Ni, Cu and the like, frequently play a harmful role because they may later diffuse into the bulk silicon and give rise to the degradation of the electrical characteristics of the devices [29]. Therefore, the analysis of metal concentrations at wafer surfaces became important in order to study metal contamination induced by fabrication processes and to check the effectiveness of cleaning processes. TXRF seems to be most promising to carry out that task since it is very sensitive, non-destructive, quick and simple to apply [30].

The removal of organic contamination from silicon or thin film surfaces is a further important cleaning task in silicon wafer processing. Organic contaminants may originate from various sources and processes: residues from photolithographic materials, residual polymers remaining from dry etching processes, organic matter from storing containers and packaging materials, crack products of vacuum pump oils, polyimide coatings and the like. Cleaning processes for the removal of organic matter principally use strongly oxidizing mixtures such as NH_3/H_2O_2, H_2SO_4/H_2O_2, HNO_3 etc. or organic solvents. Organic films such as photoresists are frequently removed by plasma etching. Apart from XPS [31], plasma chromatography mass spectrometry (PCMS) which also provides information about the nature and the quantity of the organic species [32] is now used in this field of application [28].

Layers deposited onto silicon wafers are required for isolation purposes, for interconnections, or as auxiliary materials to be removed in later steps. Deposition can be performed by several techniques like chemical vapor deposition (CVD) or sputtering. Technologically important are film thicknesses, compositions or interface reactions. These aspects can be analyzed by RBS and AES. XPS is again applied for the investigation of etching processes [33] and the testing of their selectivity [34]. For example, by means of angle-resolved XPS the thickness of passivation films can be estimated to be several nm [35]. The analysis of dopant profiles for the characterization and optimization of doping processes is the classical application of SIMS [36]. Another wide field of applications of surface analysis is failure analysis regarding yield improvement where the sample is here a more or less fully processed device.

7.4
Analytical Control of Ultra-Pure Chemicals and Chemical Processes

Ultra-pure chemicals are applied in different process steps in the semiconductor industry, whereby their influence on the success of cleaning processes aiming at the avoidance of contamination is especially important for the further steps of chip manufacturing (see Sects. 7.1.1 and 7.2) [37]. In order to avoid contamination arising from chemicals causing problems on yield and reliability in integrated circuits fabrication, the properties and detrimental impacts of the impurities have to be understood as well as their deposition mechanisms [38]. Here, surface analytical data give an idea about the origin of the impurities (see Sect. 7.3). Inorganic acids, ammonium hydroxide solution and hydrogen peroxide are used in etching and rinsing procedures while organic solvents are needed for the removal of photoresist layers from the wafer surface. The more the integration density of IC microchips (IC=integrated circuits) increases, the more the chip yields of the manufacturing processes are limited by extreme low element trace concentrations in the applied process chemicals. Therefore, tremendous demands are being made on chemical manufacturers for ultra-pure products. While in the early 1970s a maximum of 500 ng/g iron was specified, suppliers of electronic chemicals today have to guarantee maximum element trace mass ratios in the lower ng/g-range. Precise and accurate analytical procedures are therefore necessary for quality control, development, cleaning procedures and for the selection of suitable container, tubing and sealing materials [39]. With progressing development, limits of detection in the pg/g-range have to be reached.

Since up to 40 different elements have to be analyzed, preferably hyphenated techniques including sample preparation and multielement detection are applied. One of the most powerful methods for multielement analysis of process chemicals and baths with limits of detection below the ng/g threshold is ICP mass spectrometry (ICP-MS) [40]. Some elements, such as iron, calcium, potassium and silicon, are determined by means of electrothermal atomization atomic absorption spectrometry (ET-AAS) because of the better limits of determination [39]. The limits of determination obtained, for example, for silicon in acids used for etching are 1–2 ng/g.

7.5
Contamination Control of Processes and Production Installations

After a description of the investigation of silicon substrates, surface and thin film analytics as well as the necessity for analytical control of the ultra-pure chemicals and chemical processes in semiconductor technology (deposition, etching, cleaning), in which in problems of contamination and its avoidance is always regarded as most important, contamination control of processes and production installations will finally be dealt with, realizing that measurements are always the key to understanding the relationships between processes and the resulting products [41, 42].

Generating systematic data on incoming materials, processes, production environments and products by contamination monitoring and analysis is the key element of quality assurance and total quality management (TQM) [43]. Thus, contamination analysis systems are the primary tools of statistical process control (SPC) and failure mode and effects analysis (FMEA) [44], prerequisites of continuous improvement [41]. Therefore the analyst has always to invent and apply novel methods to detect traces of contaminants and their fluctuations in ever-improving process media and products. In all cases, statistical guidelines, detection and instrumental functions have to be taken into consideration. Furthermore, each analytical task begins with the evaluation of the sampling and analytical feasibility, the costs and the technological relevance of the anticipated solution to the problem.

In order to produce reliable semiconductor grade silicon, all considerable process media, process steps and intermediate product stages of wafering must be controlled by efficient analytical techniques. These analytical methods must provide reliable, relevant and real-time results. Desirably, they should be working in-line with a facile information feed-back to the process management. There are only a few procedures satisfying these requirements. A survey of the analytical methods used for process monitoring and some of their characteristics is given in Tables 7.2 and 7.3. For analytical details and characteristic data one is referred to the literature [41, 42, 45]. Generally, the atomic spectrometric methods such as AAS, ICP-OES and ICP-MS are unsuitable for direct analysis of process media. Rare exceptions are some diluted acids and ultra-pure water. The application of the methods mentioned above and also TXRF in-line applicability in surface analysis and characterization of silicon substrates have already been covered in Sects. 7.2 and 7.3.

For fast process monitoring, a couple of physical methods have been specially introduced into common practice [41], which should be mentioned here [46].

Until now, mainly inorganic analytics for wafer manufacturing have been looked at. However, organic contamination in semiconductor production are becoming increasingly important as the metallic contamination level is decreased systematically down to lower than 10^{10} atoms/cm^2. In contrast to metallic impurity control where TXRF, AAS, ICP-MS in combination with, for example, lifetime measurements are well established as state-of-the-art routine methods, there is no comparable set of equipment introduced for routine organic contamination control on wafer surfaces. The methods applied so far have been mainly used to solve singular problems, e.g. FT-IR spectrometry [47], SIMS [48], XPS [49] and GC-MS (coupling of gas chromatography and mass spectrometry) [50].

For routine monitoring, a simple method is needed which is able to analyze all possible kinds of organic contaminants in semiconductor manufacturing, such as adsorbed gases, solvents, polymers, graphite, residual oil and the like, at a sufficiently low level. In these cases, the differential thermal desorption analysis (DTDA) has been successfully applied to the control of cleaning, etching, implantation, metallization, deposition of thin films and lithographic procedures [45]. DTDA may also be used to study the desorption behavior of volatile

Table 7.2. Process and product analytical methods in the microelectronic industry

Analytcal method		AAS[a]	ICP[b]	ICP-MS[c]	IC[d]	CE[e]	TXRF[f]	NAA[g]	DTDA[h]	GFA[i]	FT-IR[j]
Media	Ultrapur water	x	x	x	x	x	x				
	Ultrapur chemicals	x	x	x	x	x					
	Process baths	x	x	x	x	x					
Process	Auxiliary materials	x	x								x
Products	Poly-silicon.	x		x	x			x	x		
	Si-ingots	x		x				x			
	Etched wafers	x		x	x	x		x	x	x	x
	Wafers: pol./epi	x		x	x	x	x	x	x		x
Materials	Packing materials			x	x			x			

[a] Atomic absorption spectrometry
[b] Inductively coupled plasma spectrometry
[c] Inductively coupled plasma/mass spectrometry
[d] Ion chromatography
[e] Capillary electrophoresis
[f] Total reflection X-ray fluorescence analysis
[g] Neutron activation analysis
[h] Differential thermal desorption analysis
[i] Gas fusion (heat extraction) analysis
[j] Fourier transform infrared spectrometry

components (solvents, monomers, moisture, hydrogen) under elevated temperatures (up to 1100 °C). It can also be applied to quantify hydrogen in silicon nitride and silicon oxide layers.

The DTDA procedure is comparatively simple; the sample is program-controlled heated in an open quartz tube from 120 °C up to 1100 °C in a constant stream of purified oxygen. The desorbing organic matter is completely oxidized at a heated catalyst to CO_2 and H_2O. These products are then analyzed by IR absorption. The intensity vs time diagram is referred to as a "spectrum" although there is no wavelength-dependent signal intensity. Hydrocarbons show coincident CO_2 and H_2O peak temperatures. In the oxidizing atmosphere hydrogen may also be analyzed. With nitrogen as carrier gas, moisture and carbonates will be detected while organic carbon is not detectable. From the combined results in oxidizing and non-oxidizing atmospheres one can discriminate between the

Table 7.3. Trace analytical capabilities of quality control in large scale Si-manufacturing

Analytical method[a]	Process/ product characteristics	Detection limit 10^{10} atoms/cm^3	Detection limit 10^{10} atoms/cm^2	Multielement method
AAS	media, auxiliary materials, wafer		0.2–3	not routinely
ICP-OES	media, auxiliary materials		less than AAS	yes
ICP-MS	media, auxiliary materials	10^2	direct: 1 after VPD: <0.1	yes less sensitive for Fe, Ca, K, Ni
IC	media, auxiliary materials		<0.2	yes
CE			<0.1	yes
TXRF	metals		polished surface: 2–30 after VPD: <0.1	yes
NAA	impurities	<1 Fe, Cu: 10^{11} at/cm^3		yes
DTDA	organic films		10^3	–
GFA	oxygen	10^6		no

[a] For explanation of the acronyms see Table 7.2

three sources of H_2O in the sample: moisture of the sample, oxidation of hydrocarbons, or oxidation of any other hydrogen-containing compound (e.g. Si-H). Although an identification of the chemical nature of the organic compounds by means of DTDA is not possible, the investigations show that many of the contaminants exhibit characteristic "spectra" and thus may be qualitatively identified especially if the history of the wafer is known. This is the case in routine monitoring for which DTDA in its present state is already suited. According to the rather simple structure of the spectra the discrimination between "good" and "bad" in routine control is easy to achieve.

References

1. Widmann D, Mader H, Friedrich H (1988) In: Heywang W, Müller R (eds) Technologie hochintegrierter Schaltungen. Halbleiter-Elektronik, Vol 19. Springer, Berlin Heidelberg New York
2. Beinvogl W (1987) ITG-Fachberichte 98:5
3. McMahon R (1991) SEMATECH Stand 3:3
4. Spanos CJ (1992) Proc IEEE 80:819

5. ISO 9000 (1987); ISO 9004 (1987) Quality management and quality assurance standards. Internat Organization for Standardization
6. Passoja DE, Casper LA, Scharman AJ (1986) ACS Symp Ser Vol 295 (Microelectron Process Inorg Mater Characterization), p 1
7. Tölg G (1987) Analyst 112:365
8. ISO 5725 (1986) Precision of test methods. Internat Organization for Standardization
9. Zulehner W (1989) In: Schulz M (ed) Landolt-Börnstein, Vol 22, Springer, Berlin Heidelberg New York, p 391
10. Kolbesen BO (1992) In: Coffa S et al. (ed) Crucial issues in semiconductor materials and processing Technologies. Kluwer Academic Publishers, p 3
11. Tölg G (1976) Naturwissenschaften 63:99
12. Woditsch P, Häßler C (1995) Nachr Chem Tech Lab 43:949
13. Bergholz B, Zoth G, Gelsdorf F, Kolbesen B O (1991) In: Bullis WM, Shimura F, Gösele U (eds) Defects in silicon-II. The Electrochem Soc, Pennington NJ, p.21
14. Kolbesen BO, Cerva H, Gelsdorf F, Zoth G, Bergholz W (1991) In: Bullis WM, Shimura F, Gösele U (eds) Defects in silicon-II. The Electrochem Soc, Pennington NJ, p 371
15. Küsters KH, Mühlhoff H M, Cerva H (1991) Nucl Instr Methods Phys Res B 55:9
16. Tanner BK, Bowen DK (1979) Characterization of crystal growth defects by X-ray methods. Plenum, New York
17. Hourai M, Murakami K, Shigematsu T, Fujino N, Shiraiwa T (1989) Jap J Appl Phys 28:2413
18. Shimura F (1989) In: Semiconductor silicon crystal technology. Academic Press, New York, p 359
19. Kolbesen BO, McCaughan DV, Vandervorst W (1990) Analytical techniques for semiconductor materials and process characterization. The Electrochem Soc, Pennington NJ
20. Eichinger P, Hage J, Huber D, Falster R (1993) In: Kolbesen BO, Claeys C, Stallhofer P, Tardif F (eds) Crystalline defects and contamination: their impact and control in device manufacturing. The Electrochem Soc, Electronics Div, Proc Vol 93–15:240
21. Prange A (1989) Spectrochim Acta 44 B:437
22. Fabry L, Pahlke S, Kotz L, Schemmel E, Berneike W (1993) In: Kolbesen BO, Claeys C, Stallhofer P, Tardif F (eds) Crystalline defects and contamination: their impact and control in device manufacturing. The Electrochem Soc, Electronics Div, Proc Vol 93–15:232
23. Penka V, Hub W (1989) Spectrochim Acta 44 B:483
24. Neumann C, Eichinger P (1991) Spectrochim Acta 46 B:1369
25. Okui Y, Yoshida M (1992) Ultraclean Technology 4:15
26. Kolbesen BO, Pamler W (1989) Fresenius' J Anal Chem 333:561
27. Riviere JC (1994) In: Ullmann's encyclopedia of industrial chemistry, vol B 6. VCH, Weinheim, p 23
28. Carr TW (1984) Plasma chromatography. Plenum Press, New York
29. Kolbesen BO, StrunKH (1985) In: Huff HR, Einspruch NG (eds) VSLI-electronics: microstructure science, vol 12: silicon materials. Academic Press, New York, p 143
30. Schwenke H, Knoth J (1982) Nucl Instrum Methods 193:239
31. Pignataro S (1995) Fresenius' J Anal Chem 353:227
32. Briggs D, Riviere JC (1983) In: Briggs D, Seah MP (eds) Practical surface analysis by Auger and X-ray photoelectron spectroscopy. Wiley, Chichester, p 87
33. Ebel MF (1978) J Electron Spectrosc Relat Phenom 14:287
34. Oehrlein GS (1986) Phys Today 39:26
35. Segner J, Mohr EG (1985) Proc 3rd Symposium Internationale sur la Gravure Seche et le Depot Plasma en Microelectronique, Cachan/Paris, p 85
36. Magee CW, Amberiadis KG (1986) In: Benninghoven A, Colton RJ, Simons DS, Werner HW (eds) Secondary ion mass spectrometry SIMS V. Springer, Berlin Heidelberg New York, p 279
37. Mertens PW, Meuris M, Schmidt HF, Verhaverbeke S, Heyns MM, Carr P, Gräf D, Schnegg A, Kubota M, Dillenbeck K, de Blank R (1993) In: Kolbesen BO, Claeys C, Stallhofer P, Tardif F (eds) Crystalline defects and contamination: their impact and control in device manufacturing. The Electrochem Soc, Electronics Div, Proc Vol 93–15:87
38. Rieger W (1993) In: Kolbesen BO, Claeys C, Stallhofer P, Tardif F (eds) Crystalline defects and contamination: their impact and control in device manufacturing. The Electrochem Soc, Electronics Div, Proc Vol 93–15:103
39. Fuchs-Pohl GR, Solinska K, Feig H (1992) Fresenius J Anal Chem 343:714
40. Tardif F, Joly JP, Lardin T, Tonti A, Patruno P, Levy D, Sievert W (1993) In: Kolbesen BO, Claeys

C, Stallhofer P, Tardif F (eds) Crystalline defects and contamination: their impact and control in device manufacturing. The Electrochem Soc, Electronics Div, Proc Vol 93–15:114

41. Fabry L, Köster L, Pahlke S, Kotz L, Hage J (1993) In: Kolbesen BO, Claeys C, Stallhofer P, Tardif F (eds) Crystalline defects and contamination: their impact and control in device manufacturing. The Electrochem Soc, Electronics Div, Proc Vol 93–15:193

42. Fabry L, Pahlke S, Kotz L, Tölg G (1994) Fresenius J Anal Chem 349:260

43. Koch KH (1996) In: Günzler H (Edit.) Accreditation and quality assurance in analytical chemistry. Springer, Berlin Heidelberg New York, p 69

44. Masing W (1994) Handbuch Qualitätsmanagement. 3. Aufl., Carl Hanser, München

45. Hub W (1993) In: Kolbesen BO, Claeys C, Stallhofer P, Tardif F (eds) Crystalline defects and contamination: their impact and control in device manufacturing. The Electrochem Soc, Electronics Div, Proc Vol 93–15:262

46. Bergholz W, Landsmann D, Schauberger P, Schoepperl B (1993) In: Kolbesen BO, Claeys C, Stallhofer P, Tardif F (eds) Crystalline defects and contamination: their impact and control in device manufacturing. The Electrochem Soc, Electronics Div, Proc Vol 93–15:69

47. Maillot P, Gondran C, Gordon M (1992) Proc Microcontamination Conf, p 681

48. de Carios J, Krusell W, McKean D, Smolinsky G, Bhat S, Doris B, Gordon M (1992) Proc Mikrocontamination Conf, p 706

49. Göbel U, Wesemann M, Bensch W, Schlögl R (1992) Fresenius J Anal Chem 343:582

50. Budde KJ, Holzapfel WJ, Beyer MM (1993) Proc 39th Annual Meeting IES, Las Vegas

8 Process Technology and Environment

8.1 Protection of the Environment and Society

The term "environmental protection" is only a few years old yet is well known and familiar to nearly everybody in the industrialized countries. This is not least due to the fact that hardly a day passes without mention of problems of protection of the environment in the daily newspapers, or on radio and television. The reason for these reports does not always seem to be the need for genuine information, but rather the "modernity" of the subject. However, it should not be overlooked that the general scale of these problems had already been recognized long before this term was coined and that there were already a lot of various activities in that field prior to that time.

Historically speaking, protection of the environment is quite an "old" subject since Hippocrates (460–377 BC) dealt with the relationship between air, water and the environment. The term "immissions" (from immittere (Latin)=to send in, introduce) first appears in ancient Rome during the reign of the Emperor Diocletian (284–305 AD) when it was a matter of reducing evil odors to a tolerable level. The recent years of the modern era show the exponentially increasing importance of this complex area which is essential for the further existence of the human species.

Special stimuli – not only in the American sphere – were emitted by US President Nixon's speech in 1970 in which he stated, among other things, "not how nature can be handled, but how we can master ourselves is the problem of these times." He thus justified his demand for a drastic reform of the protection of the environment. Here it should be remarked that – unlike the situation in the Federal Republic of Germany – in the USA at that time there was neither a law on detergents nor were there any regulations for the disposal of waste oil.

Anyone dealing with industrial processes and their control today cannot do so without looking into the relationships between industry and the environment. In many countries increasing industrialization has created conurbations in which, for a consumption-oriented society, serious problems arise as to its future. It is necessary to solve the problems arising with regard to the control of air and water pollution, the restriction of noise pollution and the disposal of waste. This enumeration shows as soon as it is looked at more closely that for the "pollution" of the environment there is no one guilty party. Everyone knows that it was only the enormous technical expansion which made possible the modern affluence of a great part of the world's population. For the valid answering of the questions an "overall" consideration of the environmental problems is necessary – that means the interaction of all the fields of life and of all the scientific and technical disciplines.

Within the scope of the overall consideration of the environmental issues, ecology is of special importance. It has to examine the relations between all living beings and their environment and thus also the relations between human living conditions and the environment. The results of that work are then to be applied in co-operation with other sciences for the improvement of environmental conditions. As certainly not all problems can be dealt with at the same time, in the awareness of collective responsibility a coordination of the tasks and projects has to take place, with priorities having to be set according to ecological points of view.

At the same time the financing of the modes of action becoming necessary must always be taken into consideration. It is surely not sufficient to develop wishful concepts without being clear about the feasibility and the cost involved. We must realize at an early stage that *we all* have to share in the financing of all these services, a fact which is all too easily overlooked. The cost of the modes of acting for the improvement of the protection of the environment, which politically and economically has to be borne by all of us, can only be met via the budgets of the municipalities, of the countries, and of the Federal Governments, respectively, and via the prices we have to pay for products required by us. Here the political aspect of the problem is shown, since the enforcement of the modes of acting for the well-being of the general public requires parliamentary action and a constant adjustment of the legal basis.

With laws and ordinances issued, a fundamental reorganization of the protection law in the sense of a protection from harmful effects on the environment and the compelling avoidance of the harmful effects has been striven for or even reached.

Preoccupation with questions about protection of the environment and the relationship between industry and environment is confirmable world-wide. The first step as a rule consists of an appraisal of the particular real situation.

As reliable statements on existing environmental conditions require exact measured results, a particularly responsible role [1] is played by chemical analytics for the study of environmental problems, since as a decision-aid analytics must create the substantial conditions for characterizing a given state or for being able to deduce the necessity of modes of acting. In addition, its methods are used for assessing and monitoring the effectiveness of all the technical processes and facilities for the disposal or avoidance of quantities of noxious matter recognized as being dangerous. Moreover, there is a reciprocal action between legal restrictions and the "effectiveness" of analytical methods.

The large variety of the analytical problems corresponds to the variety of the techniques practiced today. A list from the field of air investigation may serve as an example. Besides chromatographic separation operations and purely chemical principles, spectral photometry in various wavelength ranges, emission spectrometry, X-ray fluorescence spectral analysis, X-ray diffraction analysis as well as electron optical, electrochemical and nuclear physical methods are used. High demands are made on the determination limits of many methods which, in certain cases, are fractions of mg/kg or even ng/kg and must therefore make possible the determination of fractions of $\mu g/m^3$ or ng/m^3. Of great interest are those methods which permit continuous measurements. In some fields the share of automated methods, which are particularly important because of the virtually unlimited measuring possibilities and for the support of wide research, is more than 50%.

As in the case of all analytical tasks, here, too, the problems start with sampling. If, for example, a major air sample has to be taken in order to be able to detect very low concentrations of a substance in it, the allocation of this sample already becomes questionable. For example, an air sampling at a wind speed of 5 m/s over a time of 0.5 h means that parts which can originate from an emitting body 9 km away from the place of sampling get into the sample! Moreover, in flowing systems the knowledge of a series of parameters is important: Besides diffusion and convection processes, distribution equilibria have to be taken into account because both air and water are to be considered under the conditions of sampling as multiphase systems. Added to that is the influence of the receptacle materials in the sense of absorption, solution and chemical modification by the emission or absorption of traces. Moreover, segregation processes are to be ruled out.

The demands of protection of the environment thus always present analytics with extensive problems. For the solution of each problem an appropriate analytical management strategy [2] is required. As initially safe, i.e. low concentrations are multiplied by concentration in the vegetable or animal organism and thus, in certain cases, can possibly exceed the permissible limit, noxious matter must frequently be determinable in a concentration range over several orders of magnitude. Frequently a substance-specific, quantitative analysis is necessary, carrying out of which necessitates a method of high separation effectiveness such as gas chromatography. The determination of a-benzopyrene and e-benzopyrene which show a toxicity ratio of 1:1000 can be regarded as a problem of this type. In addition, problems can arise during the separation of solid and gaseous compounds, as in the case of fluorine, the solving of which are of direct importance for the fixing of immission limit values.

In all cases sample preparation and analysis technique (including calibration) are to be aimed in an optimum manner at a given problem [3]. This naturally applies in particular to determinations in the trace range. As round robin tests have proved inadequate, knowledge of the problem or the method can easily lead to faulty results (deviations by more than one order of magnitude) so that this results in the need to create suitable reference materials. These few examples show that the demands of protection of the environment have led to an updating of some sub-areas of analytics. The results of this discipline have made and will continue to make a considerable contribution towards gaining a complete knowledge of the natural ecological systems [4]. For these goals to be reached extensive research and development work must yet be carried out in a number of divisional fields.

It is gratifying to note that science, technology and economics attach the same importance to protection of the environment as to production. But in order to let modes of acting for protection of the environment become effective, besides the technical solution of the problems, the necessary organizational and economic conditions must be created, as protection of the environment is not the concern of a company or a branch of industry, of a town or a country, but by now a supranational, international task. For the industrial sector the world-wide handling of environmental precautions can be vital, in order, for example, to avoid competition distortions in the international market. This is because varying requirements for the design of industrial plants result in different financial burdens which then take effect as obstacles for trading between countries. The expenditure by industry on environmentally related operating costs is enormous. For example, in the German steel industry in 1987 it was already DM 50/tonne of crude steel and in the German oil refineries it rose by 1993 to approx. DM 40/tonne of crude oil. The cost of protection of the environment in the German chemical industry corresponded in 1992 to approx. 17% of net value added (BASF AG), with the expenditure on protection of the environment in the last two decades having risen 18-fold, while the net value added during the same period only grew 2.7 times.

8.2 Analysis of Air and Its Pollutants

Questions of air pollution or pollution abatement have been discussed by the general public for a long time so that it can be assumed that the problems are well known.

In the case of the problems to be handled a distinction is made between the immissions, the foreign substances in the air, and the emissions as the pollutants are discharged into the air from the most varied sources. Work is proceeding world-wide on the solution of the problems of emission limiting or even avoidance in all areas of life.

The first step towards the assessment of a detrimental effect on the environment consists of, as a rule, a determination of the actual state. Thus as early as 1970 systematic measurements were carried out in the Federal Republic of Germany with the findings that air pollution is caused 54% by motor traffic, 30% by household fires and power stations and 16% by industrial plants. Similar studies and results are known from conurbations in Switzerland and the USA. The carbon monoxide, nitrogen oxide, hydrocarbon and lead compound emissions due to motor traffic as well as the dust and SO_2 emissions of heating plants and from industry are thus in the foreground of public interest.

There is a great deal of work being done on the limiting or avoidance of dust and SO_2 emissions. Thus, for example, in the iron and steel industry, in both the blast furnace and the steelworks areas, the principle of total dust removal is more and more becoming established, i.e. the elimination of the dust sources in the unit itself and in its ancillary equipment. With awareness of international responsibility and similar appreciation of the absolute necessity for action, comparable measures were taken in other countries.

For determination and assessment, generally accepted procedures are required. The German VDI-Commission "Reinhaltung der Luft" (Air Pollution Abatement) working in the field of air pollution control since 1955, which represents one of the numerous activities of the Verein Deutscher Ingenieure (VDI) [Association of German Engineers], has been one of the main organizations to take on this task. Protection of the Environment also means an extension of the duties of the German "Technische Überwachungsvereine" (TÜV) [Technical Supervisory Associations]. From the large number of problems involved the following are selected from the area of air pollution abatement: immission protection for the approval of plants according to Art. 4 of the German Federal Immission Control Act (BImSchG), emission and immission measurements on plants of all kinds, acceptance and performance tests on gas cleaning plants, development and carrying out of large-scale measuring programs for the purpose of installing of cadastres or the calibration of continuously recording measuring devices for solid and sulfur dioxide emissions.

Exact measured results are the essential pre-condition for being able to give reliable assessments. That requires that the analyst is already involved during the test planning as only he can make the correct selection of measuring methods. For this he uses as criteria functional characteristic variables (analysis functions, calibration function, selectivity, time characteristics, ambient conditions, operating conditions), statistical characteristic variables (limit of determination, measuring uncertainty, accuracy) and operative characteristic variables (integration time, measuring range, adjustment, generation of the measuring signal, availability).

With the law for protection from harmful environmental effects by air pollution, noise, vibration and similar phenomena (Federal Immission Control Act – BImSchG) dated March 15, 1974, – since repeatedly amended – the legislator created a nation-wide basis for protecting humans as well as animals and plants, the soil, the water, the atmosphere as well as cultural and other material goods from harmful environmental influences and, to the extent that plants requiring approval are concerned, also from dangers, considerable disadvantages and substantial nuisances which are brought about in other ways and to prevent environmental influences from arising (Art. 1). The act is supplemented by various ordinances and administrative regulations in which, among other things, the determination of airborne pollutants is specified. In order to lay down and monitor emission and immission limiting values a representative and reproducible measuring technique is necessary.

The reliability of the measured results depends decisively on the problem-oriented sampling which requires the fixing of a strategy before every measurement [6]. The large variety of possible sampling methods for the constituents of air cannot be dealt with in detail at this point. Reference should be made to the literature [1, 5, 6]. The large number of substances to be determined in waste air, waste gases and in the free atmosphere corresponds to the large number of analytical methods.

The simplest form of gas detection sets is represented by the test tubes which are based on the absorption of the air pollutant to be determined on solid surfaces with the simultaneous carrying out of a color reaction [7]. With the help of a hand pump air is sucked through the test tube specific to the substance concerned and possibly adjusted to the percentage to be expected and the discoloration of which is evaluated for the evaluation of the sample. There are various manufacturers of such test tubes (e.g. Auer Gesellschaft, Thiemannstrasse1, D-12059 Berlin; Dräger Sicherheitstechnik GmbH, Revalstrasse 1, D-23560 Lübeck), which by now can be used for more than 100 test problems. The reproducibility of the measurements (relative standard deviation) fluctuates depending on the analytical task and the measuring principle between 5 and 30% [8].

Continuously measuring, automatic methods have gained particularly great importance for the analysis of air pollutants. The advantages related can be directly recognized:

1. determination and recording of pollutant concentrations without any great time delay;
2. low personnel costs;
3. avoidance of modifications in the substance of the sample by the elimination of the transport of the sample to the laboratory;
4. measurements possible at measuring places difficult of access.

The analytical methods are based on very different principles [5]. Here the colorimetry, coulometry, measurement of electric conductivity, and also physical methods based on absorption in the UV, IR and visible region can be mentioned.

In addition to these are potentiometry, galvanometry, gas chromatography and finally mass spectrometry. The basis and the fields of application of these methods have already been presented in Chap. 2 so that at this point reference can be made to it. A detailed description of the large range of measuring tasks in the field of air and waste gas analytics would go beyond the scope of this work and is also not meant to be the subject of this publication.

8.3 Analysis of the Constituents of Drinking Water, Water Fit for Industrial Use and Waste Water

8.3.1 Process Technology and Legislation

The problems with water pollution abatement are frequently considered simpler than those of air pollution abatement. That is due, on the one hand, to the fact that the volume of matter to be dealt with here is smaller and, on the other hand, to the fact that the tasks are frequently regionally more narrowly limited. But in that case, too, there are supraregional and international duties. Of the numerous modes of acting in the field of water pollution control the "International Commission for the Protection of the Rhine from Pollution" which was appointed as early as 1963 and the "International Agreement for the Prevention of Marine Pollution by Oil" can be named here.

A further difference as compared with air pollution abatement consists of the fact that water pollution abatement was assigned a special place in legislation quite a long time ago. The oldest act – almost 120 years old – is the British "River Pollution Prevention Act of 1876". For the Ruhr District the act issued by the Prussian government in 1904 deserves mention; it concerned the formation of a cooperative society for the control of the receiving water and the waste water cleaning in the Emscher area. It was characterized by two special features: responsibility for flowing waters in its whole expanse was for the first time placed under uniform administration, and administration was carried out on the basis of a co-operative society.

Subsequently a number of acts and ordinances for the protection of the waterways were issued. The nation-wide uniform rulings, e.g. inherent in the German Water Management Act and the Waste Disposal Act, are to be mentioned in particular. The water acts of the individual Länder of the Federal Republic of Germany supplement the Water Management Act and are tailored to the special problems of the Länder. Besides them the normal requirements for water cleaning plants and/or the guidelines for water dischargers designated as normal values for waste water cleaning plants are of importance as these requirements form the basis of the laying down of conditions by the authorities.

All responsible bodies are therefore aware of the fact that comprehensive water pollution control is decisive for safeguarding the future of our habitat. However, nowadays practice and everyday water management frequently do not yet appear satisfactory despite this insight. That is due not least to the considerable expenditure which is necessary, particularly in the public sector, and which cannot be implemented within the course of a few years. This group of subjects also includes the production of unobjectionable drinking water for supply to humans. Beyond the needs of each individual, drinking water also serves as an *industrial production means* as well as for the elimination of waste substances and feces, so all the questions connected with the drinking water supply are of existential importance.

Within the scope of process engineering the process materials include the drinking water used, the water fit for industrial use required and the waste water produced. Their investigation [9–11] is therefore to be seen as a part of process analytics. The various developments in the fields of water management, water preparation and treatment have led in the course of time to increased demands on water analytics.

As a result of rising water costs and increasing waste water charges it is essential to use water repeatedly and to reduce the total consumption. Optimum conditions regarding overall costs can only be reached in this situation if constant analytical control of the concentration of corrosion-inhibiting and water-stabilising chemical additives as well as of the thickening of various constituents is carried out. The possibly increased thickenings resulting from the use of organic chemicals necessitate a particularly careful analytical operational monitoring because of the associated risky technical operation mode.

In recent years waste water analysis was added to an increasing extent to these analytical tasks which are still relatively easy to handle. This partly resulted in considerably more complex work for the analyst. The stipulations regarding waste water resulting from greater environmental awareness date back more than 25 years to the creation, for example, of the German ATV work sheet [12] and the German LAWA guidelines [13] and resulted in the passing of the already mentioned Waste Water Charges Act (AbwAG) [14], of the above mentioned Water Management Act [15], of the Water Acts of the Federal Countries (LWG) [16] and of the EC Water Protection Directive [17] incorporated into German law.

For obvious reasons this resulted in the expansion of the task to be regarded in the sense of a "qualification", which to a considerable extent could be managed by optimization of the sample flow and by automation of analytical methods.

8.3.2 Water Treatment and Waste Water Analytics

The quality of waste water originating from an industrial process may be assessed from different points of view. In the course of continuously increasing requirements in the field of environmental protection with the motto "Avoidance of waste instead of disposal", water, too, must be considered a vital raw material on this premise.

In this context, analysts are required to make a statement on the question of how a special waste water may be further used as water fit for industrial purposes in other fields, perhaps after special treatment. Thus it is possible to minimize the consumption of precious drinking water according to the particular state of the art of treatment technology. Examples of this are residues from reverse osmosis plants which, depending on location and requirements, can still be used as cooling water. The same applies to heating condensates, which can be treated to produce feed-water for steam generating plants. Due to an upgrading achieved by this operation, even flushing water arranged in the correct cascade can be submitted to more effective purification, neutralization or precipitation.

If treatment is no longer economically arguable, a yardstick which, in view of continuously increasing water costs and continuously increasing waste water charges, is pointing more and more in the direction of recycling, detoxification and purification of frequently metalliferous waste water are taken into consideration as the last possible precautions. This mode of acting can be compared with the elimination of organic substances, e.g. from emulsions, so that also in this case the result, using state-of-the-art methods, will be waste water not polluting the environment. The long-term objective, however, is to reach a situation in which there are no longer any discharges of water from industrial plants into public waters.

Waste water which at present cannot be re-used under economically justifiable or technically possible conditions are subject e.g. to the Water Management Act mentioned above [15] and the North-Rhine Westphalian VGS [18] or that of the other federal countries of the Federal Republic of Germany stipulating maximum limit values of constituents for companies discharging into public sewage treatment plants. This implies that these values are constantly fixed in such a way that they correspond to the respective state-of-the-art in water clarification.

Now the analyst's task consists of controlling the observation of these values by the application of the procedures according to standards, e.g. in Germany DIN Standards, prescribed together with the limiting values (Table 8.1).

Table 8.1. Parameters for the characterization of drinking water, water fit for industrial use, and waste water

Parameter/Characteristic	German standard /DIN or 9	Analytical method
Physical and physical/ chemical characteristics		
Coloration	38404, p 1	Visual; photometry
pH-value	38404, p 5	Potentiometry
Electric conductivity	38404, p 8	Conductometry
Anions		
Chloride	38405, p 1	Argentometry; potentiometry
Bromide	D 2	Iodometry
Iodide	D 3	Iodometry
Fluoride	38405, p 4	Potentiometry
Sulfate	38405, p 5	Complexometry
Sulfite	D 6	As sulfate
Sulfide	D 7	Photometry
Nitrate	38405, p 9	Photometry
Nitrite	38405, p 10	Photometry
Phosphorus-compounds	38405, p 11	Photometry
Cyanide	38405, p 13	Volumetry
Thiosulfate	D 15	Iodometry
Thiocyanate	D 16	Photometry
Arsenic	38405, p 18	AAS[a]
Siliccic acid, dissolved	38405, p 21	Photometry
Chromate	38405, p 24	Photometry
Sulfide, dissolved	38405, p 26	Photometry
Sulfide, easily liberated	38405, p 27	Photometry
Cations		
Iron	38406, p 1, 22	Photometry; ICP-OES[b]
Manganese	38406, pp 2, 22	Photometry; ICP-OES
Calcium, magnesium	38406, pp 3, 22	AAS; ICP-OES
Hardness of water: Ca, Mg	38406, p 3	AAS; complexometry
Nitrogen (ammonium)	38406, p 5	Photometry; volumetry
Lead	38406, pp 6, 21, 22	AAS; ICP-OES
Copper	38406, pp 21, 22	Photometry; AAS; ICP-OES
Zinc	38406, pp 8, 21, 22	AAS; ICP-OES
Aluminum	38406, pp 9, 22	Photometry; ICP-OES

Table 8.1. Parameters for the characterization of drinking water, water fit for industrial use, and waste water

Parameter/Characteristic	German standard /DIN or 9	Analytical method
Cations		
Chromium	38406, pp 10, 22	AAS; ICP-OES
Nickel	38406, p 21, 22	Photometry; gravimetry; AAS; ICP-OES
Mercury	38406, p 12	AAS
Potassium	38406, p 22	Volumetry; ICP-OES
Sodium	38406, p 22	Flame emission; ICP-OES
Lithium	38406, p 22	Flame emission; ICP-OES
Silver	38406, pp 18, 21, 22	AAS; ICP-OES
Cadmium	38406, pp 19, 21, 22	AAS; ICP-OES
Vanadium	38406, p 22	Photometry; ICP-OES
Bismuth, cobalt, thallium	38406, p 21	AAS
Boron	38406, p 22	ICP-OES
Arsenic	38406, p 22	ICP-OES
Phosphorus	38406, p 22	ICP-OES
Tin	38406, p 22	ICP-OES
Combined identifiable substances		
Silicic acid	F 1	Gravimetry; colorimetry
Volatile halogenated hydrocarbons	38407, p 4	Gas chromatography
Gaseous compounds		
Oxygen, dissolved (Winkler)	38408, p 21	Iodometry
Oxygen, dissolved	38408, p 22	Amperometry
Oxygen saturation index	38408, p 23	Amperometry; iodometry
Chlorine, free and total	38408, p 4	Photometry
Hydrogen sulfide	G 3	Photometry
Cumulative parameters		
Total dry residue, filtrate dry residue, residue from glowing	38409, p 1	Gravimetry
Filterable matters, dry residue from glowing	38409, p 2	Gravimetry
Organic carbon	38409, p 3	Wet chemical oxidation, conductometry

Table 8.1. Parameters for the characterization of drinking water, water fit for industrial use, and waste water

Parameter/Characteristic	German standard /DIN or 9	Analytical method
Cumulative parameters		
Permanganate-index	38409, p 5	Volumetry
Hardness of water	38409, p 6	AAS; volumetry
Acid and base capacity	38409, p 7	Volumetry
EOX (organic halogens)	38409, p 8	Coulometry
Sedimentable matters	38409, p 9	Volumetry
Sedimentable matters	38408, p 10	Gravimetry
Nitrogen, organically bound	H 11.	Kjeldahl method
Total nitrogen	H 12	Calculation
AOX (adsorbable organic halogens)	38409, p 14	
		Coulometry
Hydrogen peroxide	38409, p 15	Photometry
Phenol index	38409, p 16	Photometry
Nonvolatile lipophilic substances	38408, p 17	Gravimetry
Hydrocarbons	38409, p 18	Photometry
Directly separable lipophilic substances	38409, p 19	Gravimetry; photometry
Organic acids (volatile in steam)	H 21	Volumetry
Exhaustible organic halogens	H 25	Coulometry
Chemical oxygen demand >15 mg/l	38408, p 41	Volumetry
Chemical oxygen demand (quick method)	38408, p 43	Volumetry
Biochemical oxygen demand	38408, p 51	Amperometry
Oxygen consumption	38408, p 52	Amperometry

Explanation:
[a] AAS = Atomic absorption spectrometry
[b] ICP-OES =Inductively coupled plasma optical emission spectrometry

Thus analytics gives, on the one hand, objective proof of the observation of environmental acts and ordinances in the water sector, and, on the other hand, it provides a basis for decisions upon the utilization chains for different types of water and aqueous solutions. Finally, it makes possible the assessment of proper operation and of adherence to system-specific parameters as far as power generating companies and water processing plants are concerned.

8.3.3 Waste Water Law and Waste Water Analytics

The demands on water analytics can only be met by the availability of instruments appropriate to industrial needs and legal regulations as well as adequately trained workers. At the same time it is indispensable for the investigation procedures to have sufficient sensitivity and limits of determination as well as low scatter to ensure an exact determination of limiting values.

The limiting values currently in force for individual elements and compounds or for cumulative parameters in the waste water sector can be seen, for example, in the last amendment of the German Water Charges Act dated 19th December 1986 [14]. Table 8.2 contains a compilation of these limiting values. They constitute examples of the close connection between analytics and statutory regulations [19].

As the charges per pollutant unit increase annually and as, despite the small concentrations of material to be charged, the calculation of loads results in considerable quantities due to the large water volumes, it becomes clear that the exact determination of pollutants is of great importance from an economic point of view.

Table 8.2. Examples of the estimation of harmful substances and cumulative parameters (according to [14])

No.	Harmful substance	Threshold value
1	COD[a]	20 mg/l; 250 kg/a
2	AOX[b]	0.1 mg/l; 10 kg/a
3	Hg	0.001 mg/l; 100 g/a
	Cd	0.005 mg/l; 500 g/a
	Cr, Ni, Pb	0.05 mg/l; 2.5 kg/a
	Cu	0.1 mg/l; 5 kg/a

[a] Chemical oxygen demand
[b] Halogenated hydrocarbons

Table 8.3. Examples of demands for the introduction of waste water of electroplating plants according to the German state-of-the-art

Parameter	Demand (mg/l)
Zn, (Sn)	2
AOX[a], Sn, sulfide	1
Pb, Cr, Cu, Ni, free Cl	0.5
Cd, cyanide (readily volatile)	0.2
Cr(VI), Ag, VHHC[b]	0.1
(BTX)[c]	(0.05)

[a] Halogenated hydrocarbons
[b] Readily volatile halogenated hydrocarbons
[c] Sum total of benzene, toluene, and xylene

Table 8.4. Demands according to the technical rules generally recognized in Germany for electroplating plants

Parameter	Demand (mg/l)
COD[a]	200
Fluoride	50
Ammonium-nitrogen	30
HC[b]	10
(Nitrite-nitrogen)	5
Al, Fe	3

[a] Chemical oxygen demand
[b] Hydrocarbons

But in order to be allowed to discharge waste water into public waters at all, administrative regulations with upper and lower limits depending on the industrial sector concerned have been decreed [20, 21], for example within the scope of the German Water Management Act (WHG) [15] forming the basis for discharge permits. A description of these circumstances the parameters and sample frequencies are shown in Tables 8.3 and 8.4 which are provided for by the 40th Administrative Regulation concerning electroplating plants.

8.3.4 Instrumental Methods in Water Analytics

After it has been shown that not only technical but especially great economic importance is attached to water and waste water analytics, it is necessary to point to the connection of progress in analysis technology and the possibilities of rationalization with the demands of modern environmental technology. In the course of the rationalization of analytical work, instrumental methods have been added to the classic chemical analysis methods.

From this point of view, for example, the use of multichannel anion analyzers for the simultaneous determination of chloride, sulfate, nitrite, nitrate, phosphate and ammonium ions (see Sect. 3.2.1.3) as well as that of ICP-spectrometry for the determination of cations in water must be seen. Further progress has been made by the use of ion chromatography for an inorganic anion analysis and those analyzers primarily based on potentiometry for the automatic determination of the electric conductivity, pH-value, acid capacities at pH 8.2 and 4.3 as well as the total amount of calcium and magnesium (water hardness).

Off-line analysis described so far is completed – especially as a contribution to rationalization – by the application of unmanned measuring stations allowing in-line measurements of waste water parameters and on-line data transmission to those departments of a company in charge of environmental problems. This continuous data acquisition probably provides the precondition necessary to meet the rigid requirements connected with the disposal of waste water. To give an example, Fig. 8.1 shows the block diagram of a waste water measuring station and explains its functions [22]:

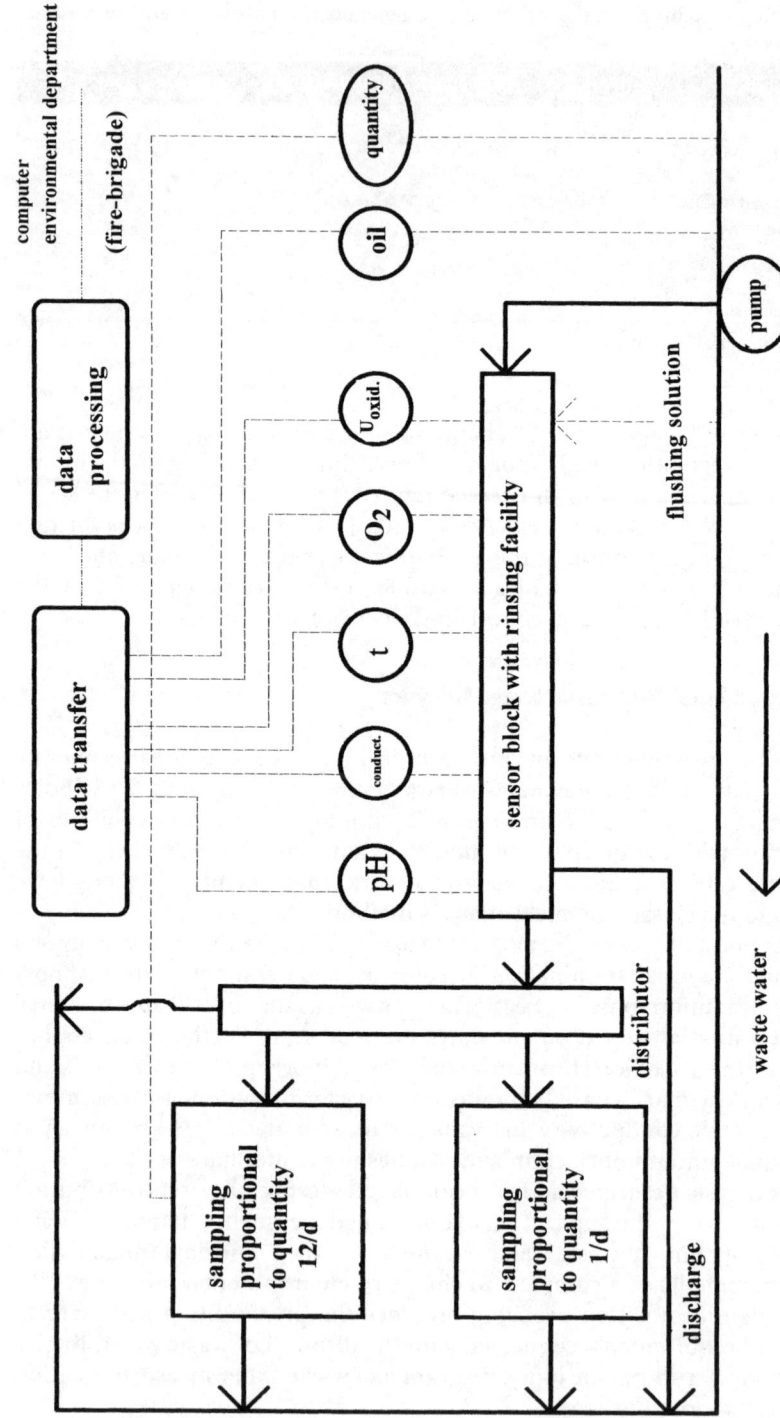

Fig. 8.1. Scheme of a measuring station for waste water control

Waste waters from various operational plants and flowing into the receiving water of a municipal sewage treatment plant are constantly monitored by means of this facility. The complete unit comprises an oil sludge tank with a continuously working sensor, which in the event of an oil leakage automatically and instantly triggers an alarm. Thus a leakage of oil into the receiving water is almost eliminated. The total amount of water flowing off over a weir of a given cross section is measured by depth measurement with echo sounding apparatus. The recorder connected to it in the waste water measuring station continuously records the flow rate of water emitted.

The waste water is in this case continuously recorded by means of sensors for the determination of conductivity, pH-value, amount of oxygen, temperature and redox potential.

At the same time quantity-proportional sample aliquots are taken off which are automatically combined to form 12 two-hour samples and one daily mixing sample. In the central chemical laboratory the daily mixing sample is analyzed for the determination of settleable solids, amount of oil, conductivity, pH-value, acid capacity at pH 8.2 and 4.3, total hardness, amounts of chloride, sulfate, iron, tin, chromium, zinc, lead, CSB, BOD-5 and TOC values. If a limiting value is exceeded, the two-hour mixing samples, which in the meantime have been stabilized by freezing, are analyzed. This procedure allows complete control of the waste water quality. The monitoring of the whole equipment is effected by a computer connected to the central alarm, so that not only the oil indicator but also any other parameter continuously determined can instantly trigger an alarm. This computer also ensures a continuous and quick flow of information between the departments involved and is designed, provided there is an undisturbed process, to reduce the great number of incoming data to a clear but representative data quantity.

Further progress has been achieved by the application of central analyzers connected to fibre-optical light guides and of chemical sensors. Here research and development is well under way in numerous fields of application. One possible solution may consist of the installation of an UV/VIS/NIR spectrophotometer with fibre-optics and light distributors, able to control any number of measuring stations. Changes in liquid fluids can be determined by means of changes in absorption or fluorescence spectra. This principle, for example, was applied to the analysis of small concentrations of chlorinated hydrocarbons and pesticides in water.

In the case of chemical sensors there is a direct interaction between analyte and sensor leading to optical and electric characteristics. These interactions must be quick and reversible. For more explanations with regard to sensor technology and, in general, concerning the analysis of liquid substances see Chap. 3 on process analytics of liquid systems.

8.3.5 Determination of Cumulative Parameters Within Water Analytics

In the field of inorganic water pollution the disturbance of aquatic or biological systems does not always depend on single ions, but mostly on their synergistic cumulative effect [23].

Thus the chemical-physical parameter of conductivity, which is easy to determine, is also a cumulative parameter making a general statement on the content of ionogenic compounds. More measurements must be added to this such as permanganate index, acid and base capacities, settleable and filterable substances, total nitrogen or hardness. The analysis procedures of these cumulative parameters must, of course, be carried out strictly according to the prescribed procedures (e.g. the German Standard Method of Water Analysis [9]) in order to lead to comparable analytical results at each of the analyzing places.

The pollution by organic material also poses problems for the analyst because of the numerous possibilities. On the other hand, the exact proportion of each single component is relatively unimportant, unless special toxins or products which cannot be easily decomposed are concerned. Their cumulative characteristics towards the aquatic system is important. Thus, e.g., it is important how much oxygen is necessary for chemical decomposition. For this purpose the cumulative parameter COD (chemical oxygen demand) was introduced in waste water analytics and included, for example, in the German Waste Water Charges Act. The same applies to BOD-5, the biological oxygen demand within a fixed period, here within five days.

Table 8.5. Examples for environmental analytical tasks in the field of solid matters

Matrix	Parameter	Analytical method
Soil	heavy metals	AAS[f] / ICP[g]
	PAH[a]	GC/MS[h]
	PCB[b]	GC/ECD[i]
	VHHC[d]	GC/MS
	BTX[c]	GC/MS
	mineral oil	IR[j]
Dust	heavy metals	AAS
	PAH	GC/MS
	PCB	GC/ECD
	CHC[e]	GC/MS
Waste	heavy metals	AAS / ICP
	PCB	GC/ECD
	PAH	GC/MS
	mineral oil	IR
	VHHC	GC/MS
	BTX	GC/MS

[a] Polycyclic aromatic hydrocarbons
[b] Polychlorinated biphenyls
[c] Sum of benzene, toluene, and xylene
[d] Readily volatile halogenated hydrocarbons
[e] Chlorinated hydrocarbons
[f] Atomic absorption spectrometry
[g] Inductively coupled plasma spectrometry
[h] Coupling gas chromatography/mass spectrometry
[i] Coupling gas chromatography/electron capture detector
[j] Infrared spectrometry

Both are important characteristics to which the cumulative parameters TOC (total organic carbon), POX (strippable, organic halogens), PCA (polycyclic aromatic compounds), phenol index, and hydrocarbons or halogenated hydrocarbons (AOX) are probably added.

8.4
Investigation of Solid Industrial Waste

Apart from the main products of chemical processes, by-products and waste are also formed. By-products are most frequently directly useable materials (for examples see Sect. 6.2.2), whereas in the case of waste there exist different procedures: recycling (e.g. metal scraps, muds, dusts), chemical transformation (combustion and energy generation) and deposition. Recycling is a principle which has been applied in industry for a long time. Examples of this are the reuse of waste produced in the production processes in the metal-working industry, the recycling of accumulators or the disintegration (shredding) of car bodies for the purpose of preparation of raw material [3]. The latest development of on-site analytics by means of spectroscopic methods (X-ray fluorescent analysis, optical emission spectrometry) provided the precondition not only for the reuse of production waste, but also the recovery of different metals of mixed scrap [24].

The examination of waste is generally carried out using the same methods known from the analytics of solids [1], only with the difference here that the analyst is often confronted with nearly insoluble problems [3]. Table 8.5 gives some examples of analytical tasks in analytics of solids in the environmental field.

The dumping of waste is subject to strict statutory regulations, and analytics provides the criteria for their observation [25]. In this context special importance has been attached to questions concerning "environmental acceptability".

References

1. Marr IL, Cresser MS, Ottendorfer LJ (1983) Environmental chemical analysis. Internat. Textbook Co, Bishopbriggs, Glasgow G64 2 NZk
2. Hein H (1991) Labor Praxis "Labor 2000–1991". Vogel Verlag und Druck KG, Würzburg, p 164
3. Leithe W (1975) Umweltschutz aus der Sicht der Chemie. Wissenschaftl. Verlagsges. mbH, Stuttgart
4. Tölg G (1993) CLB Chem Lab Biotechn 44, H 6:271
5. Leithe W (1974) Die Analyse der Luft und ihrer Verunreinigungen Wissenschaftl. Verlagsges, Stuttgart
6. Klockow D (1987) Fresenius' Z Anal Chem 326:5
7. Grosskopf K (1963) Chem Ztg 87:270
8. Leichnitz K (1967) Chem Ztg 91:141
9. Deutsche Einheitsverfahren zur Wasser-, Abwasser- und Schlammuntersuchung – Physikalische, chemische, biologische Verfahren. VCH Chemie, Weinheim
10. Fresenius W, Quentin KE, Schneider W (eds) (1988) Water analysis. Springer, Berlin Heidelberg New York
11. Tebbutt THY (1992) Principles of water quality control, 4. Pergamon/Elsevier Science, Oxford
12. ATV-Arbeitsblatt 115 (1970): Hinweise für das Einleiten von Abwässern in eine öffentliche Abwasseranlage, Regelwerk der abwassertechnischen Vereinigung e.V. (ATV) in Zusammenarbeit mit dem Verband kommunaler Städtereinigungsbetriebe (VKS)

13. Normalwerte für Abwasserreinigungsverfahren, Sonderdruck: Länderarbeitsgemeinschaft Wasser (LAWA), Hamburg 1970
14. Gesetz über Abgaben für das Einleiten von Abwasser in Gewässer (Abwasserabgabengesetz [AbwAG]) vom 13.09.1976 (BGBl.I. S. 2721, ber. S. 3007), geändert durch Gesetz vom 19.12.1986 (BGBl.I. S. 2619)
15. Gesetz zur Ordnung des Wasserhaushaltes (Wasserhaushaltsgesetz) vom 16.10.1967 (BGBl.I. S. 3017), in der Neufassung vom 25.07.1986 (BGBl.I. S. 1165)
16. Wassergesetz für das Land Nordrhein-Westfalen (Landeswassergesetz LWG) in der Fassung der Bekanntmachung vom 09.06.1989
17. Richtlinien des Rates betreffend die Verunreinigung infolge der Ableitung bestimmter gefährlicher Stoffe in die Gewässer der Gemeinschaft; Amtsblatt No L 20 vom 26.01.1980, S. 43 ff
18. Allgemeine Rahmenverwaltungsvorschrift über Mindestanforderungen an das Einleiten von Abwasser in Gewässer (Rahmen-Abwasser VwV) vom 19.12.1989 (GMBl. S. 789)
19. Hoffmann H-J (1993) LaborPraxis, "Labor 2000–1993". Vogel Verlag und Druck KG, Würzburg, p 158
20. Allgemeine Verwaltungsvorschrift über die nähere Bestimmung wassergefährdender Stoffe und ihre Einstufung entsprechend ihrer Gefährlichkeit – VwV wassergefährdende Stoffe (VwVwS); GMBl. 41 (23.03.1990), S. 114 ff
21. Frey R (1990) Metalloberfläche 44:176
22. Sebastiani E, Loose W, Koch KH (1978) Stahl und Eisen 99:1487
23. Hütter L A (1989) CLB Chem Lab Biotechn 40:64
24. Sattler H-P (1993) Materialprüf 35:312
25. Hein H (1993) LaborPraxis,"Labor 2000–1993". Vogel Verlag und Druck KG, Würzburg, p 152

9 Chemical Process Analytics as Part of Quality Management and Quality Assurance

9.1 On Quality Assurance

Archaeological findings in Jericho, the oldest industrial area in the world, point to the fact that even 10,000 years ago there was already lively trade between customers living far apart. As preconditions for the exchange of goods there were surely also in those times agreements on the quality and quantity of the products offered (mostly weapons and jewelry made of the dark mineral glass obsidian) and of the goods offered for payment (skins, wild cereals, various collected fruits, etc.). Without certain agreements co-existence and trade would have probably been threatened at that time by disputes and the use of violence. A "quality test" in that era was surely limited to only a few characteristics which could easily be observed. As early as in the Babylonian Codex Hammurabi (1750 BC) and in ancient Egyptian inscriptions, methods of quality assurance are documented. One is also reminded of the famous picture by Rembrandt (1650 AD) "De Waardijns" showing five serious-looking men who were responsible for the quality (measurements, weights) which had to be used in medieval Holland. Corporations and guilds attached great importance to high-quality work and punished offenses severely.

With technical progress sophisticated measuring methods, which have increasingly allowed a more objective assessment of materials and goods, have developed.

Today the triad quality, testing, and progress is a decisive factor when facing future challenges of the market. Therefore, it is a standing objective to optimize the combination and coordination of all quality-relevant individual corporate functions, the "quality culture". Quality management measures form an essential component of future-directed industrial activity. The market is increasingly determined by three parameters: price, quality and flexibility. The quality management of marketable products is thus becoming part of the quality policy of industrial companies.

Quality, defined as "the characteristics and features of a product or an action in their entirety to fulfill predetermined requirements" (German Standard DIN 55 350, Part 11), must be quantifiable in this context. The basis for this quantification are formed by quality inspection in the framework of quality assurance. Chemical analysis contributes to these quality ensuring measures in an important, and in many cases in an overriding and decisive way [1]. In order to meet the requirements of these tasks of product-oriented testing in the crucial field of quality policy and economy, a documented and verifiable integration of quality assurance measures into the whole analytical process is needed [2].

The changes in the sense of justice and the resulting changes of the legal situation are further reasons for these activities. The reversal of the burden of proof of the producer's product liability is concerned [3] (German Law on Product Liability of 01.01.1990) and the changed public awareness of the environment have led to the fact that questions of "quality assurance" concerning production and use of materials or machinery were presented to the public.

Quality assurance concepts cover different corporate sectors and should start at the earliest possible stage because correcting errors or deviations from the intended production target at the final inspection stage satisfies neither producer nor customer. Product and service quality are becoming increasingly significant in international competition. Thus, quality is becoming a competitive instrument for companies. However, investment in quality assurance measures is considerable because quality management systems have to be set up in a product- and company-specific way. Only in this way is it "assured" that "quality" is not an accidental result.

But as long as a production process is not mastered completely, measurements on the product are required to achieve the desired product quality. This is expensive due to test and classification processes. If it is possible, however, to eliminate the systematic causes of distributions and keep the random errors sufficiently small, a "governed" process [4] (Fig. 9.1) will result by means of quality-dependent statistical process control.

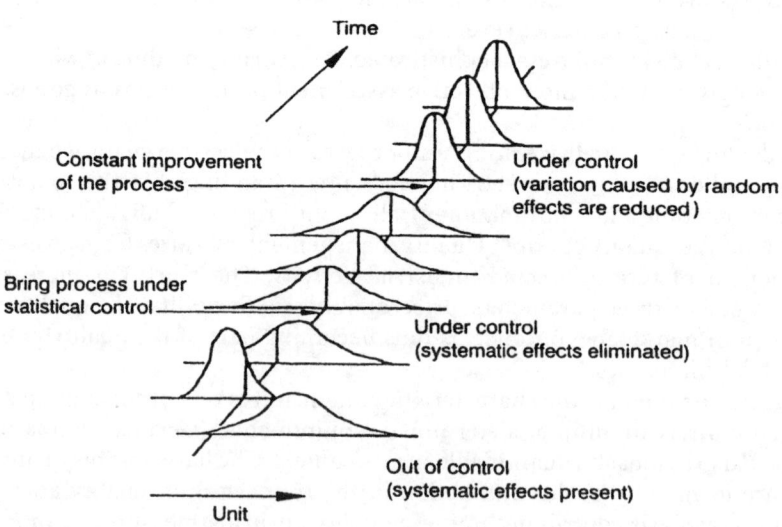

Fig. 9.1. Effect of process control

The fundamental terms of quality assurance require exact definition in order to avoid linguistic mistakes and misunderstandings; they form part of the already mentioned German standards DIN 55 350, Part 11 [5] and DIN ISO 8402 (draft). Accordingly to these, the following definitions are to be applied (partly quoted in their general meaning):

Quality management:
: all activities of the general management that determine the quality policy, the aims and responsibilities as well as their realization by means of quality planning, quality assurance and quality improvement in the framework of the quality management system.

Quality policy:
: the overall quality intentions and direction of an organization regarding quality, as formally expressed by top management.

Quality planning:
: selection, classification and weighting of the quality features as well as of a stepwise concretization of all individual requirements regarding the specifications for achieving them.

Quality control:
: preventive, surveillance and corrective measures concerning the realization of a unit (this may be results of activities and processes) aiming at meeting quality requirements.

Quality assurance:
: all planned and systematic actions which are performed within the quality management system and are laid down as required to provide confidence that a unit will satisfy the quality requirements.

Quality management system:
: the organizational structure, responsibilities, procedures, processes and required resources for the realization of quality management.

Quality inspection:
: determination of how far a unit (e.g. products or services) meets the quality requirements – or, in other words, system of testing methods, intended to meet a predetermined quality.

9.2 Quality Management Within Analytical Laboratories

9.2.1 Significance of Quality Assurance For and In Chemical Industrial Analytics

Quality management represents a comprehensive task within a company and can be classified according to functions in quality planning, quality control, and quality inspection. Within the framework of quality planning the quality features to be met are defined and weighted in consideration of the technical conditions and the customers' requirements. The quality assurance and especially its part quality inspection enable one to determine whether a product or service satisfies the quality requirements. The quality inspection is classified in test planning, test performance and data treatment. Quality control makes use of the test data resulting from the quality inspection for the purpose of controlling "quality production" and, if necessary, induces corrective measures within a running production process.

This brief characterization shows that, as far as comprehensive quality assurance is concerned, special attention must be focused on the quality inspection, as with its individual functions it comprises all the production phases of a company or branch of production. To give an example, Fig. 9.2 shows the process of quality control in steel production. At the same time it becomes clear that chemical analytics has always been of special significance in the framework of quality-assured production.

A quality-assuring and quality-assured application of testing procedures and thus the assurance of the accuracy of the test results are topics [6] which have become of increasing and fundamental importance for many divisional fields of metrology and inspection, thus including analytical chemistry.

Chemical analytics and quality assurance are linked to each other in two ways. On the one hand, analytics forms an integral part of quality assurance and provides the data needed within the framework of quality assurance to guarantee the desired properties [7]. On the other hand, these analyzed data must themselves be validated by adequate integrative quality assurance measures [8].

For some years there have been issued several international guidelines and standards in which the requirements concerning competence and acceptance of testing laboratories are laid down. In particular the following should be mentioned:

- ISO Guide 25: "Guide requirements for the technical competence of testing laboratories" and
- ISO Guide 38: "General requirements for the acceptance of testing laboratories".

The consistent complement to these guidelines is ISO Guide 49: "Guidelines for development of a Quality Manual for a testing laboratory". The scope described therein for the compiling of a quality manual includes, as a fundamental part, the demands to be made on the necessary definitions concerning the application and controlling of measuring and testing appliances (testing means) as well as recording of the accompanying data files [9].

This process of compiling quality manuals for analytical laboratories is well underway and could have already been concluded in many cases.

A further basis for working methods meeting the requirements of quality assurance is formed by the following European standards [10]:

- EN 45 001 General criteria for the operation of testing laboratories;
- EN 45 002 General criteria for the assessment of testing laboratories;
- En 45 003 General criteria for laboratory accreditation bodies.

These standards complete the standards ISO 9000 [11] and EN 29,000 which constitute the guidelines for the selection and application of standards and reg-

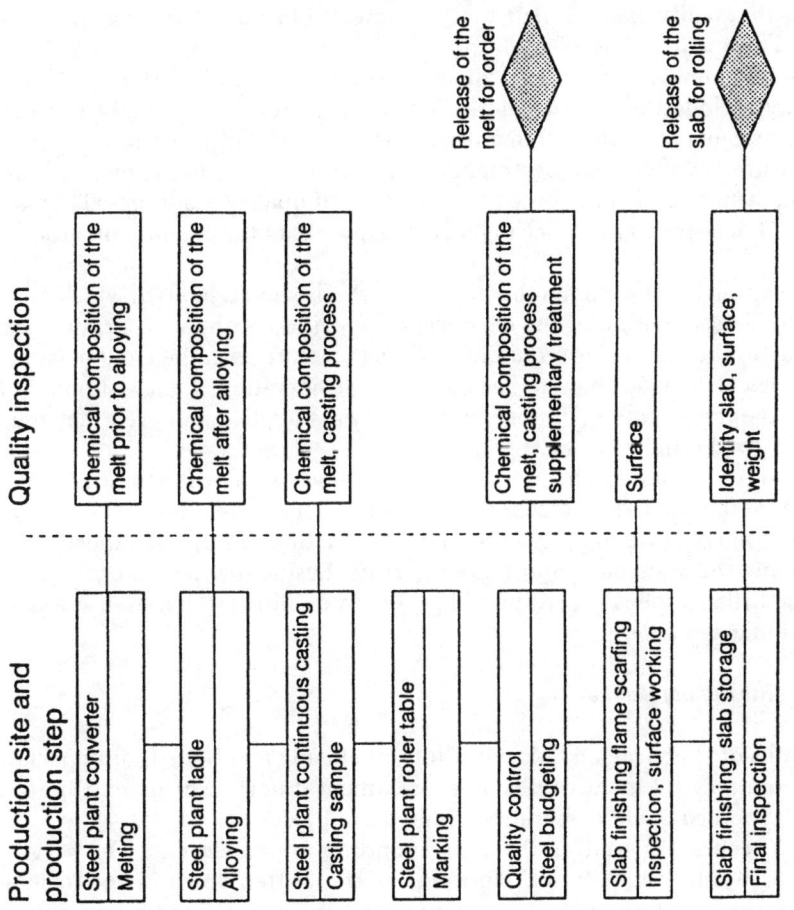

Fig. 9.2. Quality inspection for steel production

ulations for quality management, for the elements of a quality management system, and for quality assurance control levels.

In the framework of product quality assurance, chemical examination is generally included in the product-oriented measures for the quality inspection. The results of analytical investigations are, as already mentioned, part of the measured results and assessment criteria which flow into the quality management system and contribute to a coordinated quality assurance. They can even be of decisive significance in individual phases of the production processes.

This setting of tasks shows that testing methods should be used which guarantee the determination of the chemical composition with an accuracy corresponding to the particular requirements. Furthermore, in some circumstances, chronological demands have to be taken into consideration. It goes without saying that the most economic procedures are always applied, but even they must meet the given requirements.

Like all test methods, chemical analysis methods are not free of errors. It is known that the occurring total error is composed of a systematic and a random contribution [12]. Everyone involved in the production of a product must aim at minimizing the systematic and random errors. Beside the necessary technical equipment, this implies a correspondingly highly qualified staff as well as a goal-directed management.

9.2.2 The Quality Manual

The results of chemical analysis constitute, as already mentioned, an important part of the measurement results and assessment criteria which contribute to product-oriented quality assurance within a company.

The measures as well as the corresponding appliance data etc. serving quality assurance in analytical laboratories are written down in task-oriented quality manuals, the extent and content of which is defined by the above-mentioned standards and guidelines. The quality manual, which serves as an internal rule book, forms the basis of every system audit. It must be seen that all the information relevant to a system audit can be taken from it or are referred to. German standard DIN 55 350, Part 11, divides into system audit, process audit and product audit. The terms "process audit" or "product audit" are used if the effectiveness of elements of the quality management system is assessed by processes or products. "System audit" means the assessment of the effectiveness of the quality management system as a whole. "Quality audit" refers to the examination on the effectiveness of the quality management system or its elements by means of an independent, systematic inspection.

Thus the compilation of a quality manual requires adequate special knowledge about the fundamentals and requirements of quality assurance to meet these requirements. Seminars offered by various bodies are intended to fill existing gaps in knowledge concerning the field of quality assurance and to provide the necessary help for drafting of quality manuals.

A quality manual for analytical laboratories comprises the following sections, but the following subdivision and sequence should only be considered an example:

   Revision index
 0. Table of contents
 1. Quality statement
 2. Field of application
 3. Basis of the quality management system
 4. Premises and equipment
 5. Organization
 6. Staff qualifications
 7. Procurement of measuring and testing equipment
 8. Measuring and testing equipment and testing agents
 9. Checking measuring and testing equipment and testing agents
10. Test control
    – Sampling
    – Sample labeling
    – Sample transport
    – Sample preparation
    – Cross-reference to other organizational units
11. Testing
12. Quality assurance
    – Control analyses
    – Certified reference materials
    – Interlaboratory tests
    – Strategy in case of emergency
    – Internal quality audits
13. Documentation
    – Test reports
    – Updating service
14. Pertinent documents, regulations and guidelines

The terms of the titles applied for the individual sections will be briefly explained or supplemented by examples.

The revision index must provide information about the currently valid state of the quality manual showing for every individual section the currently valid version with its date of release.

Furthermore, a list of everyone possessing a numbered copy of the quality manual should be available at any time. A copy of the signed receipt for each person who possesses a quality manual is kept together with the list mentioned. The quality manual is confidential and may not be copied completely nor partially without express permission in writing given by the issuing department.

The table of contents reflects the subdivision of the quality manual and is, like all other pages, subject to the updating service.

The scope of the quality manual is defined in the introductory quality statement; furthermore, it contains a declaration of the responsible company department in which they commit themselves to observe the described quality management system. Subsequently there is an explanation of the field of application within the company and of the basis of the quality management system. The description of the field of application generally states that the introduced quality management system helps to create and maintain a high-quality standard regarding all testing activities. It is part of the quality policy of the analytical laboratory itself where it is an independent institution, or of the company of which the analytical laboratory is a divisional department.

The quality manual describes the elements of the quality management system for chemical testing and its realization in the individual fields of work. All employees are obliged to observe the rules of this manual referring to their field of activity. The head of the department assumes full responsibility for the conducting of tests in accordance with the quality manual. The quality manual contains fundamental statements on instructions for testing and documentation, and moreover the description of specific testing methods and procedures which serve to guarantee the quality of testing.

Neither should the work performed by the analytical laboratory be subject to negative influences restricting technical judgement, nor should any outside person or organization be in a position to influence the investigations and the results. Staff remuneration should depend neither on the number of tests carried out nor on their results.

The basis of the quality management system described in a quality manual is formed by the already mentioned standards ISO 9000–9004 [11], EN 29000–29004 and EN 45000 and following [10]. Furthermore, the principles of the Good Laboratory Practice (GLP Principles) [13] written down, for example, in the German hazardous chemicals legislation (Chemikaliengesetz: Gesetz zum Schutz vor gefährlichen Stoffen – ChemG of 16th September 1980, BGBl (Civil Code) I, page 1718 and following, amended on 15th September 1986, BGBl (Civil Code) I, page 1505 and following) must be considered.

Further sections cover explanations of the premises and equipment used and the organization of the department in question (flow-chart). The section on staff qualifications contains the valid quality criteria which apply to each level of organization and is complemented by documentation on job descriptions classified by field of activity and qualification certificates of further professional training and training courses.

Finally, the quality manual contains information on the documentation of the test reports and the revision service. The test results are transmitted as a written report by post, by means of telex, fax or computer-aided data transfer. These test reports are kept in the individual departments for a certain period in written form or on a data carrier. The period of safekeeping depends on the type of tested material. Details of this topic are generally included in documentation kept in the laboratory departments responsible. Every quality manual, including all related documentation and operating instructions, is subject to revision. All documents are checked once a year and, if necessary, updated. The amendments made are indicated in the revision index. The relevant documents, regulations and guidelines are listed at the end of the quality manual.

9.3 Consequences for Quality Assurance in Analytical Laboratories

9.3.1 Personnel Qualifications and Equipment

A prerequisite for the success of quality assuring measures, like the application of modern analysis technology, is qualified – and motivated – skilled personnel. The required personnel qualifications are achieved by, besides professional training, further task-oriented internal and external training; the extent of which is recorded in documents to be verified at any time. High-tech testing methods applied by either inexperienced or unqualified staff members could lead to serious problems for industry and society because of the "wrong" results obtained and the associated misinterpretation. The "testing" human is the link in a chain which reaches from the development of materials helping technical progress up to the quality-assured production of products which meet market demands.

The laboratory equipment can be assessed as being just as important as the staff's qualifications. In accordance with the laboratory's task constant effort is made to make use of the most modern technology. This leads to highly specialized analytical equipment according to the requirements of the products and to a high level of knowledge and experience of the entire staff. The organization and the equipment of the analytical laboratory should, however, not only allow an optimum handling of the order but – as already stated – should take into account the demands of quality management and possibly legal regulations (e.g. checking measurement and testing equipment, performance of control analyses etc.). This may mean for laboratories an enlargement of their fields of activity or completely new tasks, for which new methods and techniques may have to be developed in certain circumstances.

Apparatus-related progress is important in two respects for the validation of the analytical results. On the one hand, by use of the instrumental method a reduction of the number of work steps in an analysis compared with chemical procedures takes place, which automatically means a reduction of random and systematic errors. On the other hand, the use of computer-aided data processing leads to more accurate results by the avoidance of read-off, calculation and transcription errors.

9.3.2 Checking Measuring and Test Equipment

Before the procurement of an item of analysis or laboratory equipment (testing equipment) an appraisal of the suppliers concerning "quality capability" is carried out. Selection is made on the basis of the criteria applying to the particular testing equipment. That can mean, in certain circumstances, extensive analytical investigations of the testing equipment at the manufacturer based on the company's own samples.

The chemicals required for testing purposes are purchased from recognized producers on the basis of quality data. The reliability of the chemicals is constantly determined via the recording of the blank values of the analytical methods.

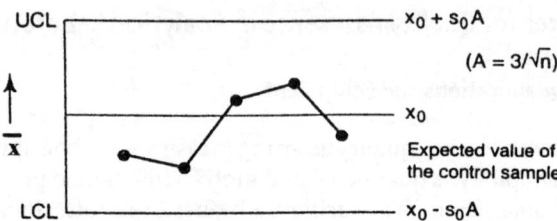

Corrective measure if:
1.) $\bar{x}$ lies outside the control limits
2.) 7 successive values of $\bar{x}$ are either located on one side of the center
    line or show continiously increasing or decreasing values

Fig. 9.3. Control chart principal

The set of problems includes the checking of measuring and test equipment, the recalibration of this equipment (calibration of measuring and test devices by means of recalibration samples), the determination of recalibration intervals and of criteria for the evaluation of analytical results in the framework of measuring and test equipment checking. The calibration state of the measuring and test equipment is regularly documented. When results are unsatisfactory, measures will be taken to remove the equipment from use and have it repaired. The results of recalibration are documented in form of control charts [14] (for example see Fig. 9.3).

9.3.3 Test Control

Test control affects sampling, sample labelling, sample handling and transport, and sample preparation. It is influenced by cross-links to other organizational units (ordering company divisions or external clients). Sampling is either carried out by the client or by the analytical laboratory. In each case, high specialized analytical knowledge is required [15]. An undisturbed work-flow may only be guaranteed if the quality manual contains unambiguous provisions and procedures for sampling, sample labelling (identification), and sample transportation.

In general, sample preparation is carried out exclusively by the analytical laboratory. It is done in keeping with the substance and the problem according to specified methods which are also described in the quality manual.

9.3.4 Test Instructions

In the framework of quality assurance within the analytical laboratory unambiguous test instructions (descriptions of the analytical procedures) should be established for all components to be determined in the materials subject to quality assurance. These instructions contain data of the equipment used and reagents needed, and an exact description of the analytical procedure including calibration and information on precision and trueness (accuracy of the mean) of the method (complete analysis instruction [16]).

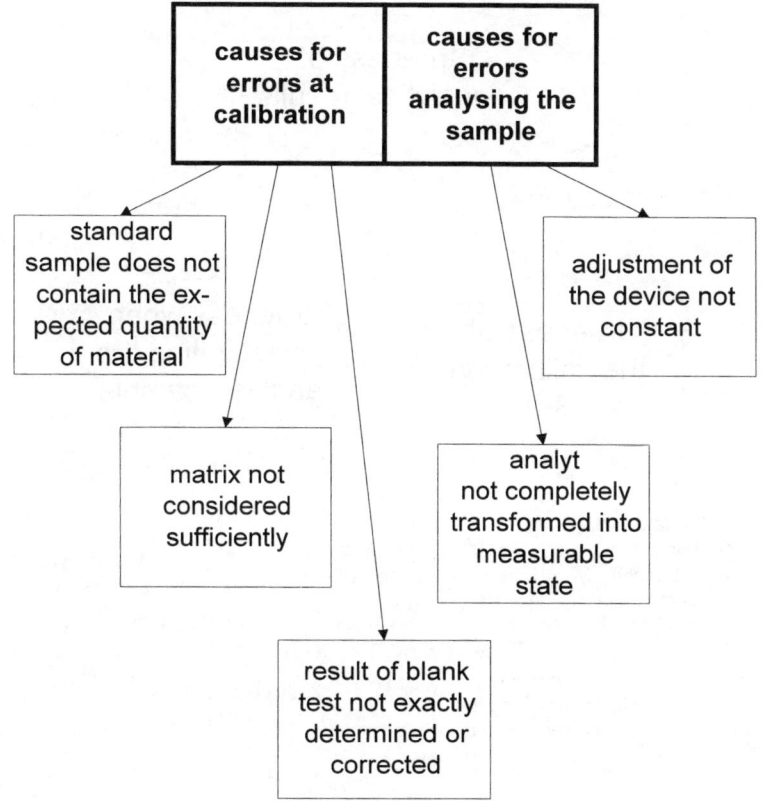

Fig. 9.4. Causes for errors in chemical analysis

For the problem-oriented application of a method, first of all, possible causes of errors must be excluded (Fig. 9.4). The trueness of the analytical results (Fig. 9.5) depends directly on the calibration of the analytical equipment which, because of the multicomponent systems generally present in the samples, cannot be performed without an exact knowledge of the chemical composition. With the use of ultra-pure substances for calibration [17] a large number of problems can be solved today. This method, however, requires adequately pure and defined substances, a demand which in the field of trace analysis cannot always be met. If one thinks of the wide field of surface analysis in this context, it becomes very clear that in this field of calibration of instrumental methods there are still a lot of open questions.

Mathematically speaking the "calibration function" is exactly described for measuring methods of all kinds (physical/mechanical or physical/chemical) as a function of the measured values depending on suitable standard values. For chemical analysis this consequently means $x = f(c)$ where x are the measured values and c is the concentration.

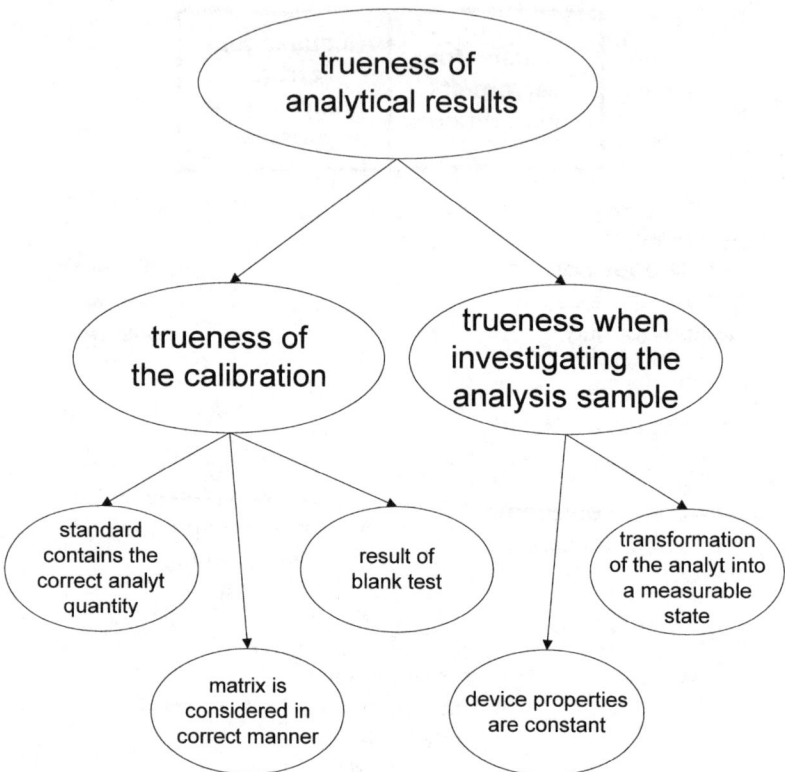

Fig. 9.5. Trueness of analytical results

H. Kaiser [18] has also described the inverse function $c = f'(x)$, by which according to his opinion he had to orient himself by the target variable when calibrating the measured variable and when analyzing the concentration, the target value. The reversal function is called analytical evaluation function. Kaiser points out that reversibility, however, is not always ensured. This is the case when there is no longer any analytical relationship, for example in the case of spectral lines with self reversal.

Calibration is done – as mentioned – by use of ultra-pure substances or by means of reference samples, of which the parameters regarding precision and trueness necessary for the analysis are known (see Sect. 9.3.6). Out of it follows the fact that reference materials are of fundamental importance for (industrial) analytics [18]. They have a twofold significance: on the one hand, they are used under certain preconditions for the setting-up of calibration functions, on the other hand, they are used as an essential means of ensuring analytical results in the sense of quality assurance.

The question of "accuracy" of chemical analysis is as old as analysis itself. However, recently it is increasingly and urgently put forward in connection with questions of standardization and quality assurance. The term "accuracy" is at first only of a very general nature and has a qualitative meaning [19]. However, wherever tests are performed, the evaluation of which is leading to a quantitative description of accuracy [20, 21], addition quantifying terms such as precision, repeatability and reproducibility as well as trueness must be used. The words accuracy and precision are used interchangeable in everyday English. Several dictionaries cite accuracy as a definition of precision (and vice versa).

As a preliminary a definition of the concept of accuracy is needed; at this point it must be emphasized that neither in the German-speaking countries nor elsewhere has consensus been reached on this question. There are an almost innumerable number of standards and an even greater number of publications on this subject without any generally recognized linguistic ruling [22] having been brought about. Therefore at the start of every new contribution dealing with questions of "accuracy" of analytical methods, definitions of the concepts used must be given in order to ensure the required comprehension.

The simplest approach is in every case to refer to standards, as they – regardless of the observation mentioned – can claim a certain general validity. According to the German Standard DIN 55 350 (Part 13) [22] by "accuracy" is to be understood "the generally qualitative designation for the approximation of assessment results to the exact or true values", where "true value" is defined as the "actual attribute value at the moment of observation under the prevailing conditions" (DIN 55 350, part 12 [23]). "Accuracy" is thus a purely qualitative, superordinated concept, while "precision" is termed as "the extent of the consistency of results as they are obtained during repeated application of a specified determination method". As expression of precision, the standard deviation of the measured values obtained under "repeat conditions" and/or the standard deviation of the measured values obtained under "comparable conditions" are generally used. The "repeatability" describes the extent of consistency between results as obtained during repeated application of a specified analytical method on an identical test object (analysis sample) at short intervals under the same conditions (the same analyst, the same apparatus, the same laboratory), the "reproducibility" correspondingly the consistency resulting at different times under different conditions (different analysts, different apparatus, different laboratories). The calculation is carried out according to the well-known mathematical/statistical rules which are described, for example, in ISO 5725 or DIN 51 848, part 3 [24].

The "trueness" (accuracy of the mean) is at first, according to DIN 55 350, part 13, the "qualitative designation for the extent of consistency between the expectation value, i.e. the mean value which was obtained by the incessantly repeated application of the determination method practiced under specified conditions and the "true value". The smaller the systematic part of the assessment deviation inherent in the analytical result the "truer" the method works. As a measure for trueness, generally used is the difference between the mean value of the result obtained according to an analytical method and the result of the "more accurate" test method ("conventional true value"). By the above-mentioned term "expectation value", according to DIN 55 350, part 21 [25], a discrete random variable is

Table 9.1. Terms for assessment and evaluation of analytical results

Term	Short definition	Standards
Accuracy	Qualitative designation for the approximation of measured results to the exact or true values $$ACCURACY:$$ $$\overline{trueness} \qquad \overline{precision}$$	DIN 55350, part 13; ISO 3534
Trueness (accuracy of the mean)	The difference between the mean value m obtained by interlaboratory comparison studies and the (conventional) true value W. $p = W - m$	DIN 55350, part 13; ISO 3534
Precision	Measure based on the standard deviation of the measured results obtained under repeatability and reproducibility conditions. The precision is quantified by the repeatability r and the reproducibility R.	DIN 55350, part 13; ISO 3534; DIN 51848, part 1,2, 3; ISO 4259, (ISO 5725)
Repeatability r	Quantitative measure for the random distribution, maximum value of the absolutely taken difference between 2 single results under repeatability conditions (one analyst, one apparatus) of normally distributed measured results (Gauss' normal distribution) with a probability of transgression of e.g. 5%, repeatability standard deviation σ_r=distribution within laboratories: $r = 1.96 \cdot \sqrt{2} \sigma_r = 2.77 \sigma_r$	
Reproducibil-ity R	Quantitative measure for the random distribution, maximum value of the absolutely taken difference between 2 results under reproducibility conditions (different analysts, different apparatus) of normally distributed measured results with a probability of occurence of e.g. 95%, reproducibility standard deviation $= \sigma_R$=distribution between laboratories: $R = 1.96 \cdot \sqrt{2} \sigma_R = 2.77 \cdot \sigma_R$ (R includes r)	

understood which results from the adding up of all measured values linked with a certain probability. A summary of these statements is shown in Table 9.1.

From the definitions mentioned above, it follows that the results have to be not only precise, i.e. should be affected with a low distribution, but also correct, i.e. without any systematic errors. A method afflicted with a (known) systematic error in certain cases provides extremely precise results, which are, however, objectively false and have to be rejected [26]. Only in the case of high precision and at the same time of a high degree of trueness can a method be described as showing great accuracy. This characteristic of measuring methods is shown in Fig. 9.6. A catalogue of measures is used for assuring the precision and the true-

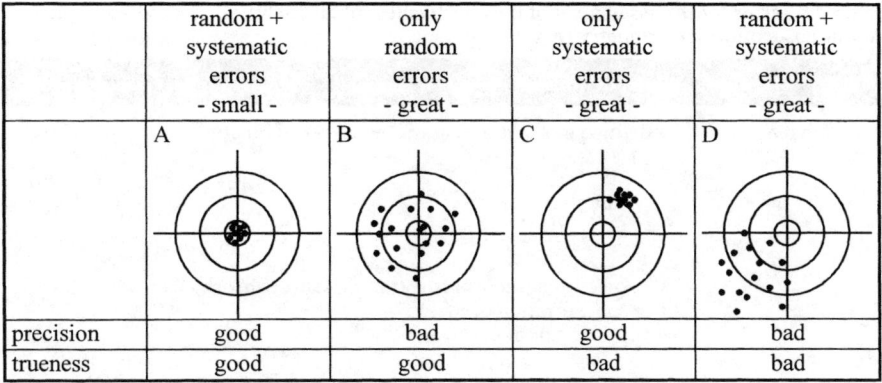

	random + systematic errors - small -	only random errors - great -	only systematic errors - great -	random + systematic errors - great -
	A	B	C	D
precision	good	bad	good	bad
trueness	good	good	bad	bad

Fig. 9.6. Precision and trueness of an analytical method

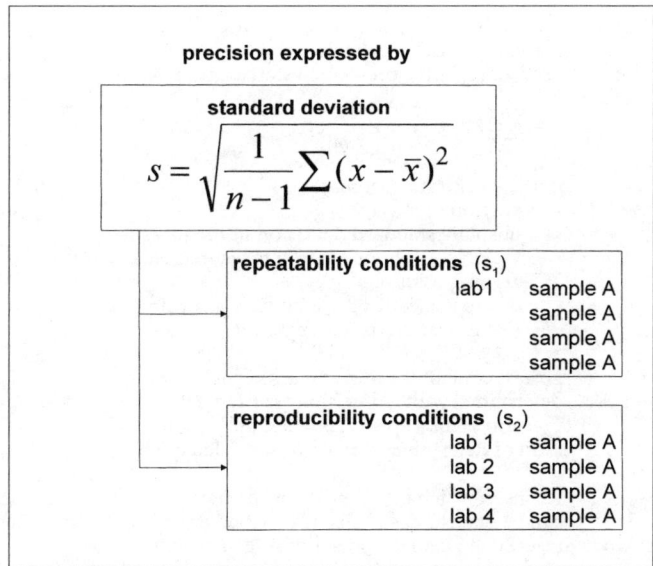

precision expressed by

standard deviation

$$s = \sqrt{\frac{1}{n-1}\sum(x-\bar{x})^2}$$

repeatability conditions (s_1)
lab1   sample A
sample A
sample A
sample A

reproducibility conditions (s_2)
lab 1   sample A
lab 2   sample A
lab 3   sample A
lab 4   sample A

Fig. 9.7. Calculation of the standard deviation under repeatability and reproducibility conditions

ness of analytical results. From the application of quality assurance measures the problem-oriented selection and control of the analytical methods, the use of recognized reference materials and the performance of interlaboratory comparison tests are primarily to be mentioned.

Nearly all statistical assessments of measured results are based on the determination of the standard deviation. It is a parameter for the characterization of the distribution of results, i.e. of the precision of measured values (Fig. 9.7). However, the stating of a standard deviation or of variables derived from it only

Table 9.2. Terms for the assessment of the effectiveness of an analytical method: example – atomic absorption spectrometry (AAS)

Term	Short definition	Standard
Sensitivity	Is defined by the gradient of the reference function $A = f(\beta)$ e.g. $\dfrac{\sigma A}{\sigma \beta}$ or $\dfrac{\sigma A}{\sigma W}$ A = spectral decadic absorption measure=absorbance β = concentration of mass w = portion of mass	DIN 1319, part 2, DIN 51401, part 1
Characteristic concentration (or mass)	Concentration (or mass) of the element to be determined which is measured with an absorbance A=0.0044 (1% absorption factor)	DIN 51401, part 1
Limit of detection	The limit of detection indicates the concentration (or mass) which can be detected by a defined procedure with a prescribed statistical certainty, e.g. $\beta = \dfrac{\delta\beta}{\delta A} \cdot k \cdot s$ or $w = \dfrac{\delta w}{\delta A} \cdot k \cdot s$ β = concentration of mass w = portion of mass s = absolute standard deviation of the measuring variable A (obtained from measurements of zeroing solutions) k = factor, usually 2 or 3, is chosen according to the desired statistical certainty	
Limit of determination	Lowest content (or lowest mass) which can be determined with a precision (see Table 9.1) prescribed for a special case of application. limit of determination=3 × limit of detection	DIN 51401, part 1
Analytical concentration range (or mass range)	Range in which the content (or the mass) can be determined with a prescribed precision (see Table 9.1). The lower limiting value of the analytical concentration range (or mass range) is the limit of determination	DIN 51401, part 1

becomes complete and thus meaningful when the test conditions are also stated. Thereby it can be understood immediately that the standard deviation in the case of repeated testing of a sample in a laboratory – here designated as s_1 – is smaller than the standard deviation s_2 which reflects the distribution of results of various laboratories.

If the standard deviation of an analytical method has been determined by numerous tests, from this value various variables can be calculated which provide meaningful information regarding analytical practice. For example, within a range of ±2 s and in the case of normal distribution, 95% of all the measured values are to be expected if the measurement is repeated for an arbitrary number of times. Furthermore, the admissible deviations R1 and R2 are often determined according to the American Standard ASTM E 173–68. The R1 and R2 values roughly correspond to about three times the standard deviation s_1 or s_2. These characteristic values have the following meanings: R1 is, according to the standard mentioned, the admissible deviation between two values which were determined of the same sample within one laboratory, and R2 is the admissible deviation of the values which were determined by two different laboratories.

The application of outlier tests cannot be dealt with at this point. Here reference must be made to the (extensive) literature [16, 27] and existing standards (see References to Chap. 9).

Finally, it remains to be mentioned that methods to be applied to a particular problem must have sufficient sensitivity and an appropriate limit of determination so that the conditions are provided for clear determinability (Table 9.2).

9.3.5 Control Analyses

Special attention must be paid within analytical quality assurance measures to the assuring of test results (validation) [28, 29]. The measures include, besides the minimization of the analytical errors by technical developments and staff training, the production of reference materials and the participation in interlaboratory comparison tests, the requirement to perform control analyses regularly. A first "safety measure" of all analytical studies is to perform duplicate determinations with a subsequent statistical evaluation of the results. If the individual values deviate from one another more than expected for the technique applied, "control analyses" are performed. If inadmissible and not directly explainable deviations occur, then, in certain circumstances, the use of this measuring or test equipment will be suspended pending further investigation.

A further measure for validating the analytical data is to perform control analyses on samples chosen at random and according to well-determined test procedures and/or recognized reference procedures [30]. This means an increase in expenditure; however, this should be considered necessary in relation to the possible harm from not doing it. The results of these control analyses are documented in the form of quality control charts (Fig. 9.3).

In this context it is certainly of interest to get an estimate of the expenditure needed for quality assurance. The total expenditure within the laboratories for the validation of analytical results and their control differs from one laboratory to another. However, it can be assumed that it amounts to about 30% of the total laboratory expenditure in time and costs. From this fact it automatically follows that analyses of comparable quality can only be obtained by the application of comparable systems and by the existence of corresponding experience.

In connection with the quality-assured application of analytical methods attention should be drawn to the value of standards which can become important for the assuring of analytical results by means of the comparison of methods or the application of standardized reference methods. In the case of demands resulting, for example, from producers' liability it is always an advantage to use or to observe standards as they contain the recognized rules of technology [31]. The non-observance of the relevant standards can, in a case of damage, be classed as "proof of first appearance" that the damage was culpably caused. From that the conclusion must be clearly drawn that the progress of standardization projects in the whole field of industrial testing technology, including also chemical analytics, must not only be followed with interest but has also to be actively influenced by the divisional analytical bodies since it leads finally to the issuing of rules which are a pre-requisite for unequivocal and frictionless operating sequences [32].

9.3.6 Reference Materials

For the assuring of analytical results, reference materials (RM) or certified reference materials (CRM) are of fundamental importance (see e.g. DIN 32 811) [33]. Certified reference materials (CRM), i.e. materials with confirmed amounts of

Table 9.3. Example for a certificate of a reference material: boron containing alloy steel (laboratory means of selected elements based on four determinations in % m/m)

Lab. No.	C	Cr	Ni	B
1	0.0148	18.48	10.24	0.8291
2	0.0156	18.48	10.24	0.8375
3	0.0157	18.50	10.25	0.8399
4	0.0157	18.52	10.26	0.8470
5	0.0158	18.52	10.31	0.8575
6	0.0158	18.56	10.32	0.8729
7	0.0162	18.57	10.32	0.8778
8	0.0162	18.61	10.32	0.8870
9	0.0163	18.62	10.34	0.9050
10	0.0166	18.62	10.36	0.9102
11	0.0167	18.63	10.36	0.9273
12	0.0168	18.63	10.38	0.9283
13	0.0171	18.63	10.39	0.9302
14	0.0173	18.64	10.40	0.9344
15	0.0178	18.65	10.40	0.9375
16	0.0179	18.66	10.40	0.9400
17	–	18.67	10.42	–
18	–	18.74	10.44	–
19	–	18.77	10.44	–
MM[a]	0.0164	18.61	10.35	0.8914
sM[b]	0.0008	0.08	0.07	0.0401

[a] Arithmetic mean of the laboratory means
[b] Mean standard deviation of the laboratory means

Certified values of selected elements with respect to the statistical evaluation of the individual values (in% m/m)

Lab. No.	C	Cr	Ni	B
MM	0.016	18.61	10.35	0.89
sM	0.001	0.08	0.07	0.04

substance, are produced and distributed by internationally recognized organizations or institutions [34].

According to the German standard DIN 32 811, by reference material a "material or a substance of which one or more properties are so precisely specified that they can be used for the re-calibration of measuring instruments and the control of the results of measuring, testing and analytical methods as well as for the characterization of material properties" is understood. According to the type of chemical specifications a distinction is made between primary reference materials (ultra-pure substances or standard titrimetric substances), of which the parameters are mainly determined by the chemical formula and secondary reference materials, the parameters of which are specified by investigation according to recognized methods. Generally speaking a reference material makes possible the transfer of the value of a specified property between various places or points in time. It can consist of a gas, a liquid, a solid substance, but also of a simple manufactured item. Table 9.3 shows the certificate of a CRM which not only gives information on the chemical composition of the material but also statistical information on the admissible distribution of the analytical data and the laboratories involved in certification.

In analytical laboratories, according to the cited standard, reference materials are used for calibration ("calibration standards") and for controlling ("control samples for analyses"). However, the certified data alone do not guarantee the successful, i.e. correct application of the reference materials. Depending on the material to be investigated or on the testing method to be applied, expert assessment and problem-related selection is required. It follows that the task-related application of an (instrumental) analytical method including the calibration standards still demands the professionally trained specialist.

9.3.7 Interlaboratory Studies

Interlaboratory comparison tests (round robin analyses) are performed in order to characterize the quality of analytical methods, e.g. within the framework of standardization, and during the preparation process of certified reference materials. The planning and performance of these studies is subject to certain rules [35] the results of which serve as an assessment basis after statistical evaluation [36]. These co-operative investigations are also a necessary procedure to secure quality assurance in analytical laboratories [37].

It is the aim of these cooperations that participants gain insight into their own performance and into the comparability of the analytical results. Each participant can deduce for himself from the results of these interlaboratory tests the suitability of the applied analytical method, his staff's professionalism, and the reliability of his equipment, thus proving his competence.

9.3.8 Quality Audits

Internal quality audits are performed with the aim of determining the actual state of the quality management system and of the course of quality assurance measures, and to attain continuous progress and improvement. Therefore, the quality management system is audited at least every two years. The working basis for the audits are the quality manual together with the relevant operation manuals, records and documentation.

Internal quality audits are planned and prepared by the person responsible for the quality management of the company department. However, they are performed by auditors not belonging to the department to be audited. Planning and working out the program for the audit is done with the assistance of specialists of the analytical laboratory. The auditors compile a final report on the result of the audit including possible corrective measures that have been agreed upon. Carrying out the scheduled corrective and improvement measures is overseen and reviewed by repeated audits if necessary.

9.4 Process Capability and Machine Capability

For the statistical evaluation of production and measurement processes the terms "process capability" and "machine capability" play a central role. Process capability means that the production process (analytical process) constantly meets the quality requirements, while machine capability provides information as to whether a machine or a procedure is operational with recognizable uniformity within preset tolerance limits.

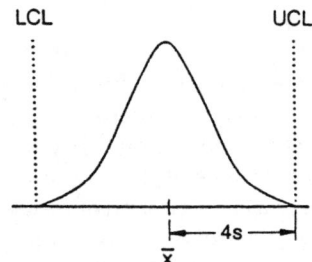

- only random variation
- normally distributed

LCL   UCL

$\longmapsto 4s \longrightarrow$

$\bar{x}$

Minimal requirement: $\bar{x} \pm 4s$ specification = 99.994 %

$$\Longrightarrow \quad c_m = \frac{UCL - LCL}{6s} \geq 1.33$$

resp. considering the mean value position:

$$c_{mk} = \min \left(\frac{UCL - \bar{x}}{3s} \; ; \; \frac{\bar{x} - LCL}{3s} \right) \geq 1.33$$

Fig. 9.8. Calculation of machine capability

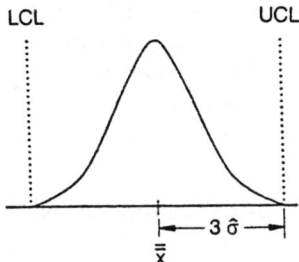

- long-term study
- cooperation of personnel, machinery,
  raw material, methods and work environment

Minimal requirement: $\bar{\bar{x}} \pm 3\,\hat{\sigma}$ specification
$= 99.73\ \%$

$$\Longrightarrow \quad \boxed{c_p = \frac{UCL - LCL}{6\,\hat{\sigma}} \geq 1.0}$$

$= $ estimate $\hat{\sigma} = \dfrac{\bar{R}}{d_2}$ or $\hat{\sigma} = \dfrac{\bar{s}}{c_4}$

d_2, c_4 are depending on the sampling size.
Process coefficient:

$$\boxed{c_{pk} = \min\left(\frac{UCL - \bar{\bar{x}}}{3\,\hat{\sigma}} \; ; \; \frac{\bar{\bar{x}} - LCL}{3\,\hat{\sigma}}\right) \geq 1.0}$$

Fig. 9.9. Calculation of process capability

Process capability is an index number which indicates whether a process runs within required limits. The variation and the accuracy of analytical results exert a significant influence on this index number which should not be less than 1. Improvements in process capability as well as in machine capability (>1.3), can be achieved by smaller standard deviations and by the elimination of systematic errors. Reporting the capability indices with time is an appropriate means of proving and assuring analytical quality and, therefore, it also contributes to ensuring the quality of processes related to analysis [38].

While machine capability is a short-term observation, evidence of process capability requires a long-term study. In both cases, the variation of the (analytical) method, expressed as standard deviation, play an important role in the calculation of the indices (see Figs. 9.8 and 9.9) [39, 40]. Here, the approach deviates from the pure analytical process as the variation should be in appropriate relation to a tolerance that is determined to be the result of the technical process by defining an upper and lower limit. The process capability index indicates whether it is possible to run the specific process in the required way.

In this context, the mean value of the measured test parameter becomes particularly important. The process is only operable if this mean value equals a target over a given period of time. Machine capability determines the performance of a machine or a measuring instrument with regard to its keeping within defined tolerance limits. In this case, requirements are higher because a variation of 4 s (99.994% of all test values) is used for the calculation. Optimum performance of the machine or of the measuring instrument has to be proven by performing a great number of measurements over a short period. Systematic errors must be eliminated prior to study process capability and machine capability. With regard to the valuation and significance of these value figures one must refer to the literature.

9.5 Accreditation of Analytical Laboratories

At home and abroad, an increasing number of purchasers of industrial products expect their suppliers not only to run an effective quality management system but also to have system effectiveness proved by a neutral body. Therefore, in the interests of industry and commerce, organizations which issue such certificates were founded. Besides the certifying of quality management systems a need has arisen for the accreditation of testing laboratories, i.e. for analytical laboratories, too. As a basis of these activities there exists the already mentioned European series of standards EN 45,000, according to which accreditation of a testing laboratory means the formal recognition of the competence of a test laboratory to carry out certain tests or types of tests [41].

To obtain accreditation, this is, of course, a free decision of the individual laboratory since there is no stringent governmental guideline. The market alone is the determining factor. Many laboratories have, however, recognized the increase in competitiveness accreditation gives and take part in the procedure in order to demonstrate their analytical quality.

The accreditation of testing laboratories in the area not subject to statutory rulings is further accompanied by a considerable economic aspect. Assuming that in the future all testing laboratories within the European Community are accredited by their national accreditation bodies following the same criteria (EN 45,002), then no compelling reason exists any longer to carry out the same test on product conformity at the customer's site. That means (theoretically) saving half of the test costs.

Advantages resulting from the accreditation of an analytical laboratory can be summarized as follows:

1. qualified fulfillment of customer expectations for the effectiveness of the quality management system;
2. a stronger position when auditing a company's quality management system;
3. harmonization of commercial arbitration analyses (here, for example, a fundamental difference exists between mechanical/technological and metallurgical testing);
4. proven competence with regard to the execution of assignments from outside the company;
5. qualified participation in interlaboratory studies on the production of certified reference materials (CRM).

In the latter case it is the best and most economic way for self control and an examination of one's own performance. Moreover, it provides the possibility of introducing one's own competence into the CRM procedure and exerting an influence on CRM developments.

It is not possible to deal at this point with the complex multi-level co-operation of all the different organizations working in the field of certification and accreditation. For that purpose reference has to be made to the corresponding literature [42].

References

1. Koch KH (1990) Kontrolle, Oktober:80
2. Günzler H (ed) (1996) Accreditation and quality assurance in analytical chemistry. Springer, Berlin Heidelberg New York
3. Theuer A (1990) LABO 12/1990:7
4. Kremer K-J (1989) Stahl u. Eisen 109:119
5. DIN 55350 (Part 11) (1987): Concepts in quality and statistics; basic concepts in quality management, quality control and quality assurance
6. Thierig D, Thiemann E (1983) Arch Eisenhüttenwes 54:301
7. Czabon V (1992) Fresenius J Anal Chem 342:301
8. Hartmann E (1992) Fresenius J Anal Chem 342:764
9. Qualitätssicherungshandbuch und Verfahrensanweisungen DGQ-Schrift Nr. 12- 62, 2. Aufl., 1991, Deutsche Gesellschaft für Qualität e.V. (DGQ), Frankfurt/M.; Beuth Verlag GmbH, Berlin
10. Böshagen U (1996) The accreditation of chemical laboratories. In: Günzler H (ed) (1996) Accreditation and quality assurance in analytical chemistry. Springer, Berlin Heidelberg New York, p 15
11. ISO 9000: Guidelines for selection and use of the standards on quality management, quality system elements and quality assurance; ISO 9001: Quality systems – Model for quality assurance in design/development, production, installation and servicing; ISO 9002: Quality systems – Model for quality assurance in production and installation; ISO 9003: Quality systems – Model for quality assurance in final inspection and test; ISO 9004: Quality management and quality system elements – Guidelines
12. Danzer K (1996) Significance of statistics in quality assurance. In: Günzler H (ed) (1996) Accreditation and quality assurance in analytical chemistry. Springer, Berlin Heidelberg New York, p 105
13. Merz W, Weberruß U, Wittlinger R (1992) Fresenius`J. Anal Chem 342:779
14. Werner W (1992) Fresenius J Anal Chem 342:783
15. Wegscheider W (1996) Proper sampling: a precondition for accurate analyses. In: Günzler H (ed) (1996) Accreditation and quality assurance in analytical chemistry. Springer, Berlin Heidelberg New York, p 95
16. Doerffel K, Eckschlager K (1981) Optimale Strategien in der Analytik. H. Deutsch, Thun, Frankfurt/M
17. Danzer K (1995) Fresenius J Anal Chem 351:3
18. Griepink B (1990) Fresenius J Anal Chem 338:360
19. Adams MJ (1995) Chemometrics in analytical spectroscopy". The Royal Society of Chemistry, Cambridge
20. Handbuch für das Eisenhüttenlaboratorium, Bd. 5, Kap 6.6 Angewandte Statistik für Laboratorien, Stahleisen, Düsseldorf, 1985
21. Farrant TJ (1997) Practical statistics for the analytical scientist. The Royal Society of Chemistry, Cambridge
22. DIN 55350, Part 13 (1987): Concepts in quality and statistics; concepts relating to the accuracy of methods of determination and of results of determination
23. DIN 55350, Part 12 (1987): Concepts in quality and statistics; attribute-related concepts
24. DIN 51848, Part 3 (1972): Precision of test methods; calculation of values for the repeatability and reproducibility

25. DIN 55350, Part 21 ((1987): Concepts in quality and statistics; random values and probability distributions
26. Koch KH (1986) CLB Chem Lab Betr 37:282
27. Gottschalk G, Kaiser KE (1976) Einführung in die Varianzanalyse und Ringversuche. B-I-Hochschultaschenbücher, Bd. 775
28. Ebel S (1992) Fresenius J Anal Chem 342:769
29. Wegscheider W (1996) Validation of analytical methods. In: Günzler H (ed) (1996) Accreditation and quality assurance in analytical chemistry. Springer, Berlin Heidelberg New York, p 135
30. Koch KH (1982) Arch. Eisenhüttenwes. 53:97
31. Budde E (1979) DIN-Mitt. 58:351
32. Fremgens GJ (1979) DIN-Mitt. 58:603
33. Griepink B, Quevauviller Ph (1996) Reference materials for quality assurance. In: Günzler H (ed) (1996) Accreditation and quality assurance in analytical chemistry. Springer, Berlin Heidelberg New York, p 195
34. De Bievre P (1996) Traceability of measurements. In: Günzler H (ed) (1996) Accreditation and quality assurance in analytical chemistry. Springer, Berlin Heidelberg New York, p 159
35. Cofino WP (1996) Accreditation and interlaboratory studies. In: Günzler H (ed) (1996) Accreditation and quality assurance in analytical chemistry. Springer, Berlin Heidelberg New York, p 209
36. ISO 5725: Precision of test methods; determination of repeatability and reproducibility for a Standard Test Method by interlaboratory tests
37. Grasserbauer M, Pfannhauser W, Wegscheider W (1987) ÖChemZ 88:130
38. Thierig D (1991) Stahl u. Eisen 111(10):83
39. Ford AG (1985) Köln: Statistische Prozeßregelung, Qual. Cont. EV 880 b
40. Ford AG (1988) Köln: Q 101, Qualitätssystem. Richtlinie
41. Staats G, Tröbs V (1992) CLB Chem. Lab. Biotechn. 43, H. 4:194; H. 6:314
42. Günzler H (1996) EURACHEM – Organization for the promotion of quality assurance in analytical chemistry and the accreditation of analytical laboratories in Europe. In: Günzler H (ed) (1996) Accreditation and quality assurance in analytical chemistry. Springer, Berlin Heidelberg New York, p 241

10 Economy of Industrial Analytics

The economy or profitability of industrial analytics must be looked at from two points of view. On the one hand, it plays as to its effect a considerable or even decisive and in special cases quantifiable role in the industrial process field; in any case, it contributes to the economy of technical processes. On the other hand, it must orient itself to economic criteria [1] (cf. the dual problem circle: automation with and within analytical chemistry!). The possibilities for analytics to influence the economy of technical production processes and the questions related with quality assurance of products can be directly deduced from the presentation of the process analytical methods serving the various branches of industry and from the demands of quality assurance. The minimization of the costs striven for in every process, seen against the background of progressive technical development and the commercial pressures constantly acting on the production of industrial products, is a permanent demand; however, the fulfillment succeeds only for a limited period.

An essential prerequisite for economic acting in analytics [2] is the limiting of the analytical order to the extent absolutely essential for the technical process in question. That necessitates constant dialogue between the customers and the analyst in order to define in each individual case the necessary number of samples for a given problem, the number of the components to be determined and the tolerable scatter of the analytical results. Zettler [3] presents the costs as a function of standard deviations and obtains the relationship

$$N = \left(\frac{t \cdot s}{\pm U}\right)^2 \approx costs$$

where

N is necessary number and/or size of samples,
t is factor from the STUDENT distribution,
s is standard deviations of material, sampling and pre-analysis,
± U is tolerated uncertainty of the result.

It represents an estimate of a number or size of random samples for the sampling to be carried out. On the other hand, it means that a cost increase is the consequence of the rising standard deviations of material, sampling and analysis and/or decreasing value for the tolerated analytical results or the uncertainty specified by other demands.

The occasionally published formulae on the subject of "minimization of laboratory costs" [4, 5] should not be dealt with in detail here as they are not very informative or helpful for the respective case to be considered.

Questions of the economy of analytical methods should be looked at very carefully when automation of analytical procedures is involved. The commercial benefit of quick (automated) test methods, adequate for the process, for production and product control can be summarized by the following points:

1. reduction of the number of misfit melts or batches;
2. increase in the product quality;
3. decrease of the specific direct costs by high capacity of the units due to short times for analysis;
4. reduction of the costs of basic and auxiliary materials by increase of the precision of the analytical methods;
5. decrease of the specific analysis costs and the increase in productivity of the laboratory.

The decision concerning the use of "instrumental" analytical methods with a correspondingly high investment expenditure as replacement for conventional "manual" works practice assumes a well-founded cost comparison. In such economy calculations, the capital to be invested, the lifetime of the plant, the annual operation costs and the personnel costs to be saved as well as other savings or additional expenditure on material costs have to be included. As an example for such a comparison, results of calculations from the field of spectrometric metal analytics can be mentioned [6]. According to them it could be shown that the ratio of the spectrometric analysis costs to the number of analyses can be described by a hyperbola (Fig. 10.1). By comparison, the ratio of the chemical analyses costs to the number of analyses remains constant, as with an increase of the number, which would result in an increase of man-power and a proportional increase in material consumption, the costs per analysis does not change. With an increase in the number of spectrometric analyses no further manpower is required and no proportional additional consumption of energy and auxiliaries occurs.

In the example given the costs of chemical analysis was taken as being 100%. From a certain number the costs of spectrometric analysis is lower than that carried out chemically. On the other hand, it becomes clear that the costs of spectrometric analysis of a low number of samples as a result of the higher capital service and the higher expenditure on maintenance and repairs can be considerably higher than that of the chemical analysis. In this example, the costs are equal at 60,000 analyses per annum. Here, it has not yet been taken into account that in the case of further elements to be determined the costs of the chemical analysis increases independently while the costs of the spectrometric analysis (almost) does not alter. As a function of the degree of automatization of an analytical system the investment to be raised increases exponentially (Fig. 10.2) so that decisions concerning the supply of capital for (further) increasing the degree of automatization require careful economic calculations.

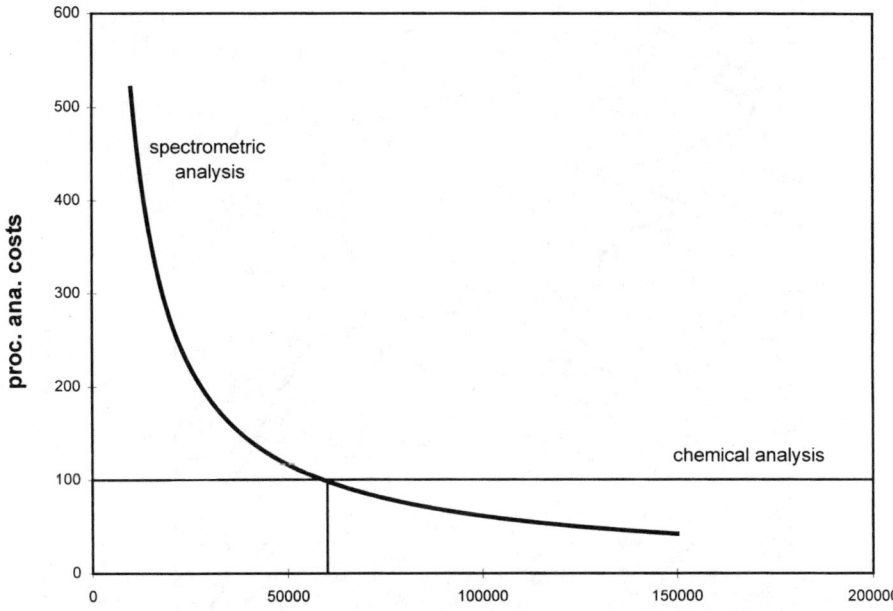

Fig. 10.1. Comparison of costs between chemical and spectrometric analysis of metals (example)

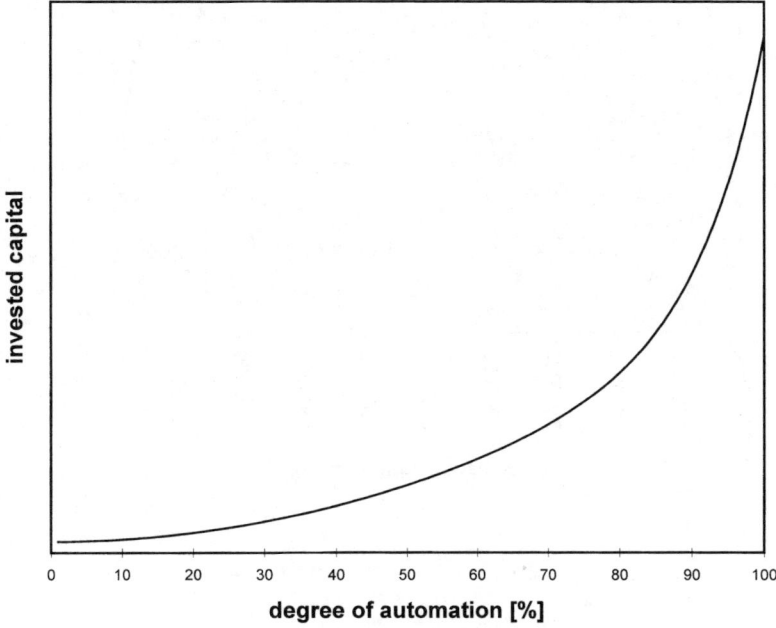

Fig. 10.2. Relation between investment and degree of automation (schematically)

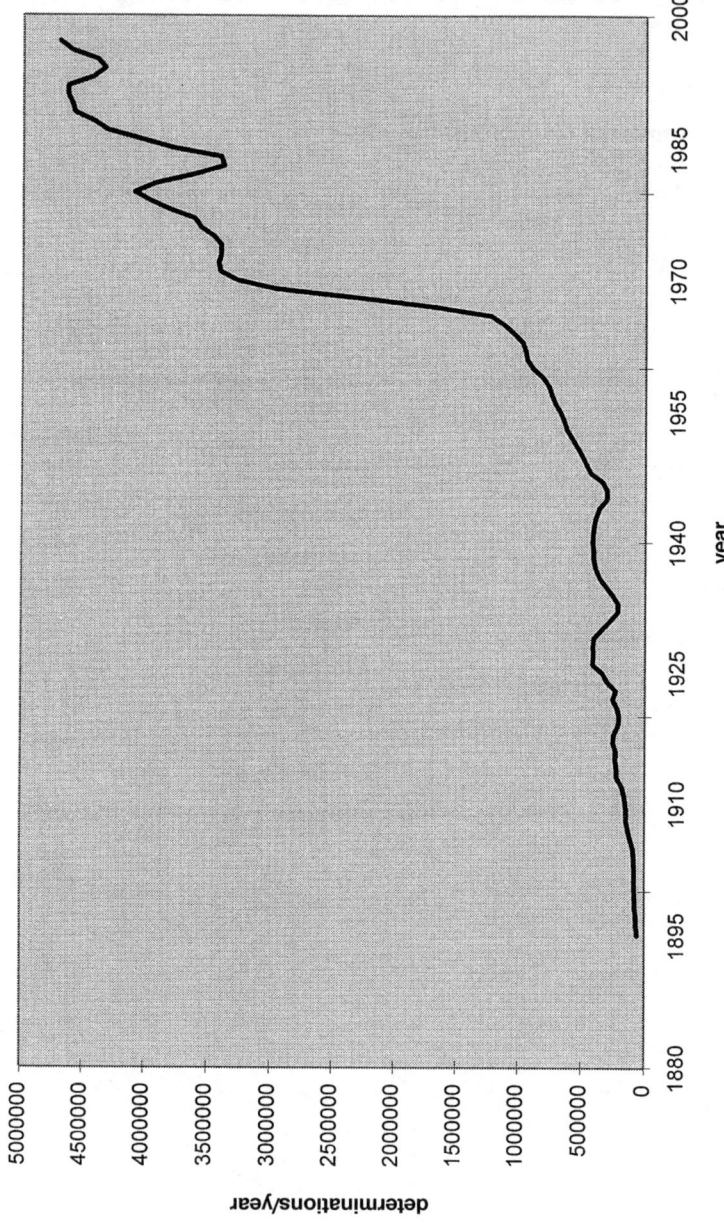

Fig. 10.3. Increase of analytical determinations of a steels works laboratory in historic view (example)

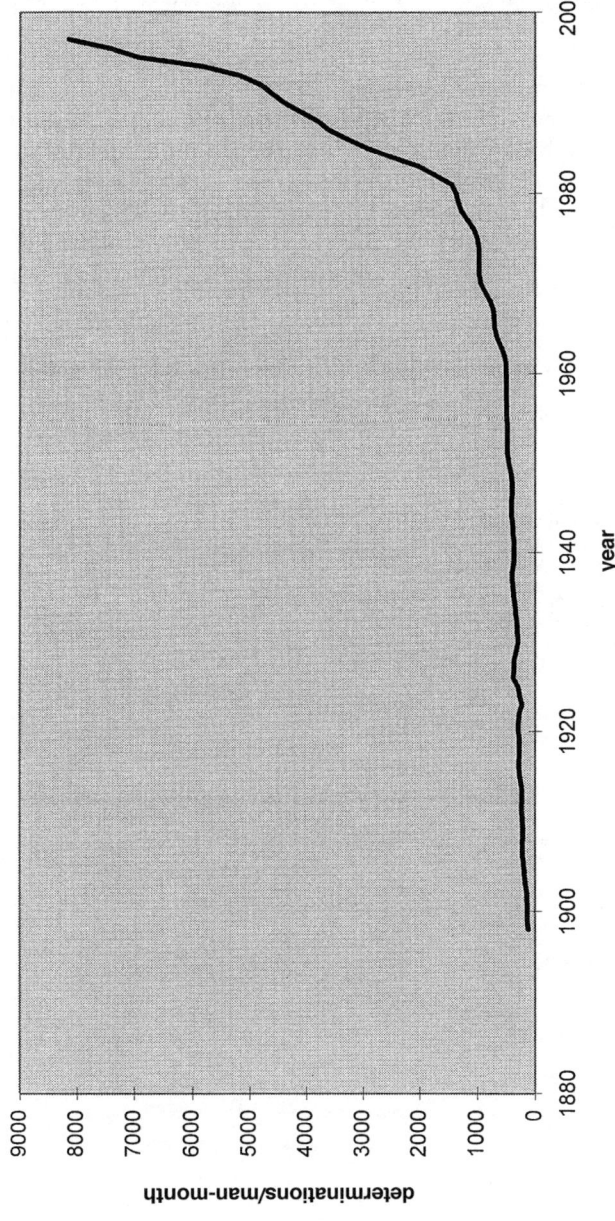

Fig. 10.4. Development of the "analytical productivity" of a steel works laboratory since 1890 (example)

The economic success of the use of instrumental methods in process analytics can be explained by the following figures from the steel industry [7]. In chemical analytics a gradual change took place analogously to technical development which is characterized by the increasing use of analytical equipment up to automatized "analysis lines". On the one hand, this resulted in a constant rise in the total number of analytical investigations in the laboratories (Fig. 10.3) and, on the other hand, in an exponential rise in the "analytical productivity", expressed as determinations/analyst and month, (Fig. 10.4). Of course, this development cannot be continued indefinitely: this is shown by the right-hand curve ends which at the present time are slowly tending towards a horizontal line.

References

1. Scharrnbeck C (1970) Chem Techn 22:166
2. Doerffel K, Eckschlager K (1981) Optimale Strategien in der Analytik. H. Deutsch, Frankfurt/M, p 39
3. Zettler H (1978) Erzmetall 31:460
4. Vandeginste BGM, Janse TAHM (1977) Fresenius' Z Anal Chem 286:327
5. Fahr E (1977) Fresenius' Z Anal Chem 287:97
6. Kromidas S, Guardiola J (1994) LaborPraxis — Mai:34
7. Gegner H, Kunze D (1968) Neue Hütte 13:308
8. Koch KH (1990) Fresenius 'J. Anal Chem 337:229

11 Outlook on Research and Development Trends in Industrial Process and Product Analytics

The limits of technical systems are mostly determined by the effectiveness of the materials used (available). Research and development in the field of new products are therefore considered priority tasks just like the optimization of well-known materials. Material research thus means, like cooperation in the optimization of technical processes a great challenge for analytical chemistry. Here, the methods of process analytics in the narrower sense are to be further developed, and, for example, in materials sciences material analytics with its large variety of methods of micro, micro-area, trace and surface analysis is challenged. At the same time, the same applies for chemical analytics as it is stated for the products. There must be and will be new developments as well as the optimization of well-known methods.

The statements in one of the foregoing chapters have shown the current state of applied atomic spectroscopy. After a few decades of development it has become a sector of considerable process technical and economic importance, and today its instrumental development still cannot be considered fully concluded. This statement is also valid to the same extent for many other methodical developments. Due to the economic situation in wide areas of industry, however, a change in practice-related development work can be observed. In earlier years, a large share of application work was carried out in industrial analytical laboratories closely oriented to daily routine. Nowadays, these tasks have shifted more and more to the manufacturers of (spectral) analytical equipment from whom now individual analytical solutions are expected. They often also have to provide client-specific software.

Process technical, material technological, economic, and also environmental formulations of questions will in the future be the motive spring for further developments, where the following fields having to be considered as the main areas in the sense dealt with in this book.

11.1 Process Analytics and Automation

In earlier chapters it was explained in detail that optimization of processes and the quality assurance demands powerful process analytics. Here, future development goes to on-line and in-line methods which can be integrated into process control systems. Non-destructive working optical and spectrometric methods will have a high share in this field.

In the case of product analytics the demands of quality assurance will lead to a clearly increased number of samples, the testing of which with constant or even reduced personnel in the analytical laboratories can only be overcome by consistent automation. Hence, it follow there will be demands for ever more effective, more flexible automatic sample preparation and free-programmable robot systems as well as minimal user expense by more pretentious computer or microprocessor operation. Thereby, new automatic device operating software works to an increasing extent with standardized user surfaces.

Beyond that, the market more often demands the networked computer as a requirement for placing orders, for example by means of a long-distance data transfer or for the electronic transmission of analytical certificates or safety data sheets of products delivered or to be delivered. Summing up, it can be ascertained that electronic data processing will become a key function. For only in this way can the immense (analytical) data quantities be documented appropriately for inquiries to be made accessible and for statistical evaluations, for example for the investigation and valuation of a production process, to be used.

The application of chemical sensors will play an increasingly more important role for the control of production processes besides the installation of personnel-free operated laboratory units as well as automatic analytical systems "on site" (at-line) [1]. There are, of course, critical voices which say that some market prognoses are too euphoric for chemical sensors, but we can be sure that further development of this technology will continue market-oriented and successfully. That is valid particularly for electrochemical, thermocatalytic, ion conducting and ionization sensors. But also the development of chemically sensitized field effect transistors, of fiber optical and piezoelectric sensors will lead to interesting process analytical applications [2]. A special trend to miniaturized sensor systems can be observed.

However, it should be mentioned here that in chemical sensorics the crucial points for decision criteria for new and subsequent development of sensors will be determined above all in that way; advantages will be achieved over competing "conventional" analytical methods and user interest in improvements in the usual analytical practices will be vigorous. Only a detailed market and need analysis can give information about what will be the areas of activity.

For future process analytics the concept of Total Analysis Systems (TAS) developed by Widmer and coworkers [3] will attain especially noticeable importance because there are no limitations as in the field of sensorics (see Sect. 5.3.2). These are flexibly applicable, complete analytical systems with which proved analytical principles, such as high performance liquid chromatography (HPLC), capillary electrophoresis (CE), and flow injection analysis (FIA) are used as process bases. The photolithographic manufacture of microflow systems enables the miniaturization of the last-named technology (μ-TAS) and opens up thereby additional fields of application for the FIA (see Sect. 5.3.3). Future developments aim at the integration of detectors and pumps as well as the creation of microscale systems. Additional possibilities for microscale analytical systems are HPLC and CE, of which the separating column or the separating capillary, respectively are etched into a silicon chip (chip-technology). Hence, it can be assumed that these micromethods will, in the course of the next few years, play a significant role in the analysis of extremely small sample volumes as on-line measurement in "Total Analysis Systems".

The advantages which these analytical systems offer as fast and selective analyses with low consumption of chemicals and low yield of waste materials, recommend them for on-line process analytics and will open great opportunities on the analytical market [3].

11.2 Spectroscopic Method Developments

In process and product analytics the spectroscopic methods have – as repeatedly mentioned – always played an important role. Thus at this point a short glance should be cast at some actual trends of development. Of course, there have hardly been fundamental changes in the last few years regarding, for example, UV-/VIS-spectrometers; however there has been increased development towards simultaneously detecting diode array spectrometers. Further development in microsystem technology together with monolithic components will lead to low-priced integrated optical systems, which will find their application in optical sensor technology and in in-line process analytics.

Analytical chemistry faces great challenges regarding the direct analysis of solid matters, of which new materials such as high performance ceramics, multilayer materials, surface treated and composite materials are of great technological and economic interest. The development of electrically non-conducting materials has accelerated methods which use laser evaporation and radio frequency sputtering of the sample material for atomic spectrometric analysis. Here glow discharge spectrometry still finds numerous fields of application and needs further development, for example regarding the investigation of non-metals and non-metallic coatings by means of high frequency lamps and the procedures for calibration in the case of depth profile as well as of micro-distribution analysis. Up to now, reference materials are rare for these purposes.

The last-mentioned problem is of fundamental significance for analytics and will be an important task for analytical research for a long time. Finally, the reliability of analytical methods depends on the production of sufficient and problem-oriented certified reference materials, characterized as much as possible by competent institutions. Here, a task exists to be solved in a community of experts, which can never be considered completed because of constantly progressing analytical development.

Important fields of instrumental innovation affect systems of on-line sample preparation and sample feeding for atomic spectrometric methods. As primary radiation sources for atomic spectroscopy, diode lasers are developed. Moreover, laser spectroscopy can be expected to develop as a method for significant applications in the near future, particularly for micro and trace analysis of technical products [4].

A few years ago, near-infrared (NIR-) spectrometry was seldom applied for process and product analytical purposes. Due to the development of better spectrometers, the use of fast computers and the application of chemometric methods, it is today often the procedure of choice for rapid identity tests, for example of stock receipt or as part of product control. Surely the effectiveness of this analytical technique is not yet dealt with completely. Broader applications may be expected not only in the food industry but also in the pharmaceutical industry and for testing of polymer foils [5].

With this example the outlook on spectroscopic method developments can be closed. It should be remembered that in each chapter of this book references are to be found to promising and forward-looking research processes and developments.

11.3 Material Sciences

Progress in material sciences are closely tied to progress in process and product analytics. That is not only valid for the classic metals and alloys, but to a higher degree for the "new" materials such as high performance ceramics, surface coated and composite materials. The progress to be achieved with these materials depends directly on the state of analytics including microscopic methods. The general research target is seen today combining microscopic properties with macroscopic characteristics of a material and understanding completely new materials as to their structure and their behavior. As a further characteristic of this future-oriented field, which can be regarded as a challenge for both industry and science, technological progress in this area can only be gained by a tight interdisciplinary cooperation of solid matter chemists and physicists as well as materials scientists and analysts.

The analytical questions in the case of ceramic materials affect the base materials, such as Al_2O_3, AlN, Si_3N_4, SiC, as well as the ceramic final products [6]. For analysis of the base materials ICP and atomic absorption spectrometry as well as classic methods have been applied. The analysis of the final products is far more difficult, because in this case not only the bulk composition but to a deciding extent the microdistribution of the elements and the structure must be of interest. For solving the analytical problems mainly spectroscopic methods, such as secondary neutral particle and ICP mass spectrometry as well as different laser spectroscopic methods, are applied [7]. Moreover, for the characterization of non-metallic materials X-ray diffraction, infrared and Raman spectroscopy and increasingly solid-NMR-spectroscopy are used. In all methods there exist open questions with regard to reliable applicability, which is the object of current analytical research as well as production of reference materials for this extremely complex area regarding the assurance of the results of investigation.

For the investigation of phase transformations at gas/solid reactions, raster (atomic) force microscopy can be used [8]. Highly dispersing electron microscopy proves a suitable means for the investigation of structural properties as well as for bonding relationships and atomic distances at internal boundary layers. Thereby possible statements about microstructural conditions can be brought up for the description of macroscopic characteristics.

The rapid progress of "high-tech-material-sciences" always puts forward new demands to the analytical chemist and accelerates the development of new procedures and methods. Strong impulses, for example for the preparation of surface analytical methods, emerged from the development of microelectronics, optoelectronics, super hard coatings and special polymers (for example for biomaterials) [9]. Today, a variety of methods are available for the analysis of surfaces and intermediate layers, of which only some should be mentioned here. Electron probe microanalysis (EPMA) for characterizing the surface morphology and for microanalysis, analytical electron microscopy (AEM) for highly dispersed interface analysis, Auger electron spectroscopy (AES) for depth profile and microsurface analysis, X-ray photoelectron spectroscopy (XPS) for the investigation of chemical bondings in surface layers, secondary ion mass spectrometry (SIMS) for trace analysis in surfaces and intermediate layers as well as atomic force microscopy (AFM) for the topographical surface characterization with atomic resolution [10]. In spite of the fact that all these methods have reached a high standard, a lot of problem-oriented research work still has to be done [9].

The analysis of solid matters differs from those of gases and liquids by another factor – topographical and structural analysis. There is no solid material that does not depend on the amount of secondary and trace components. It therefore follows that there is a need for methods with high detection sensitivity for multi-trace analysis in order to be able to investigate the influence of lowest impurities on the structure and the properties of materials. Of similar importance is the distribution and micro-area analysis of trace elements within the material matrix and at grain boundaries or in intermediate layers. These methods enable statements about the influence of trace elements and element distribution on material properties [11]. Here, analytics not only has to create the presuppositions for new future-oriented technological developments but also has to provide means for a reliable quality assurance when producing these new materials.

References

1. Göpel W, Hesse, J, Zemel JN (1991) Sensors – a comprehensive survey, part 1: chemical sensors. VCH, Weinheim
2. Carey WP (1994) TrAC 13:210
3. Manz A, Verpoorte E, Raymond DE, Effenhauser CS, Burggraf N, Widmer HM (1995) In: Van den Berg A, Bergfeld P (eds) Micro total analysis systems. Kluwer, p 5
4. Niemax K (1991) Naturwissenschaften 78:250
5. Benson IB (1995) Spectroscopy Europe 7(6):18
6. Broekaert JAC, Tölg G (1990) Microchim Acta II:173
7. Niemax K (1990) Fresenius J Anal Chem 337:551
8. Friedbacher G (1995) Nachr Chem Tech Lab 43(3):342
9. Grasserbauer M, Friedbacher G, Hutter H, Stingeder G (1993) Fresenius J Anal Chem 346:594
10. Grasserbauer M, Werner HW (1991) Analysis of microelectronic materials and devices. Wiley, Chichester
11. Ortner HM, Wilhartitz P (1991) Mikrochim Acta II:177

12 Subject Index

Springer
and the
environment

At Springer we firmly believe that an international science publisher has a special obligation to the environment, and our corporate policies consistently reflect this conviction.

We also expect our business partners – paper mills, printers, packaging manufacturers, etc. – to commit themselves to using materials and production processes that do not harm the environment. The paper in this book is made from low- or no-chlorine pulp and is acid free, in conformance with international standards for paper permanency.

Springer

Printing (computer to plate): Mercedes-Druck, Berlin
Binding: Buchbinderei Lüderitz & Bauer, Berlin